Sensors and Measurement Techniques
for Chemical Gas Lasers

Mainuddin
Gaurav Singhal
A. L. Dawar

Sensors and Measurement Techniques
for Chemical Gas Lasers

IFSA International Frequency Sensor Association Publishing, S. L.

Mainuddin, Gaurav Singhal, A. L. Dawar
Sensors and Measurement Techniques for Chemical Gas Lasers

ISBN-13: 978-84-617-1152-9
ISBN-10: 84-617-1152-1
BN-20140711-XX
BIC: TJFC

Contents

Preface

During the last few decades, Lasers have emerged as the most innovative tool, having wide ranging applications starting from the very common supermarket bar code readers to the highly advanced systems such as nuclear fusion systems for power generation, directed energy weapon in an antimissile role. Lasers are well established in many manufacturing technologies, for precision delivery of intense power for scribing, cutting, welding, and for precise 2D and 3D metrology. There is hardly any sphere of our life, which is untouched by Lasers. A common man encounters Lasers not only in light shows, but also with the beauticians, eye specialists, orthopedists, to sight a few applications. In industry lasers are being widely used for material processing, alignment etc. Doctors use it for variety of applications like surgery, dermatology, dentistry, ophthalmology etc. The use of lasers by medical community, industry and by the academic community continues to increase day after day. Many educational institutions are using a wide variety of lasers on regular basis for demonstration of various experiments. These devices produce intense beam of coherent light that can be concentrated to the point where considerable energy densities are required. These can either be pulsed lasers or CW lasers and can produce visible to infrared wavelengths. Pulsed lasers can have energies from microjoules to kilojoules, whereas the CW lasers can have powers ranging from milliwatts to Megawatts.

Gas lasers such as Carbon dioxide gas dynamic laser (CO_2 GDL, λ =10.6 µm), Hydrogen fluoride-Deuterium fluoride (HF-DF, λ=2.7-3.4 µm), and Chemical Oxygen Iodine Laser (COIL, λ =1.315 µm) etc. are infrared gas lasers having wide range of applications in various defense and industrial scenarios. However, in chemical laser the population inversion is based on direct or indirect exothermic chemical reaction occurring in the cavity itself. Polanyi initially suggested the chemical laser concept in 1961. But it was Kasper and Pimentel at university of California, who experimentally realized the emission, for the first time in 1965. The main attraction of chemical lasers is their ability to produce very high powers of the order of megawatts. Common examples of chemical lasers are the Chemical Oxygen Iodine Laser (COIL), Hydrogen Fluoride Laser and Deuterium Fluoride Lasers. When such laser devices are driven by chemical reactions, the power of the laser beam can be spectacular. However, the development of chemical lasers involves many scientific and technical challenges in multidisciplinary branches of science including chemistry, physics, fluid dynamics, optics, and engineering. These devices are now finding important applications in industrial and military operations. HF / DF lasers of megawatt class have already proved their Directed energy weapon potential in the form of Mid-Infrared Advanced Chemical Laser (MIRACL) and Tactical High Energy Laser (THEL). Directed energy weapons based on COIL are also being developed, such as Boeing's Airborne Laser which has been constructed inside a Boeing 747 and is intended to destroy short- and intermediate-range ballistic missiles in their boost phase.

Sensing and Measurement is the key technology area in the development of these lasers. Advanced sensing and measurement technologies are required to acquire, analyze and transform data into information that is useful to enhance the performance and capabilities of these lasers systems.

Beginning with a brief introduction to the basic concepts of sensors and transducers, classification of various sensors, the monograph discusses characteristics of various sensors, sensor signals, signal conditioning and their operations in chemical lasers scenarios. The emphasis is laid more on applicable diagnostic techniques, data processing and hardware selection. Further, design methodology and implementation of customized data acquisition system required for gathering and analyzing information in real time regarding various critical parameters vital for optimal system performance has also been detailed.

First Chapter gives a brief introduction to Chemical Lasers and highlights the importance of sensors and measurement of various performance parameters for extracting optimal performance.

Second Chapter highlights the sensors for measurement of basic parameters like temperature, pressure, photon detection and flow metering and control. The selection of these sensors for chemical lasers applications have also been discussed in this chapter.

Advanced and customized diagnostic techniques for measurement of various derived parameters such as specie concentration of gaseous fuels and intermediate species, cavity medium characterizations in terms of small signal gain, Mach number, medium homogeneity etc. have been discussed in chapter three. These are essential for providing useful insight into the complex phenomenon occurring inside the laser cavity enabling their optimization.

Both direct sensors and diagnostics setups require signal-conditioning for transmitting data in a form suitable for data acquisition system. Chapter four discusses this aspect for various kinds of sensors employed in chemical gas lasers.

Chapter five briefly discusses various aspects of Data Acquisition system (DAS) and sensor interfacing issues with it. Since chemical laser operation is for a short duration of the order of few seconds only, all the data has to be captured and processed during this period. This chapter also discusses the selection of various components of data acquisition system. The implementation of data acquisition from point of view of operation, analysis and safety, a critical area of concern, has also been discussed. The role of data acquisition system in laser optimization and concurrence with theoretical aspects has also been illustrated by a few examples.

Chapter 6 finally gives an insight into the uncertainty analysis in the measurement of various parameters mentioned in chapter two and three.

The thrust is essentially on the description of basic sensors, related hardware, techniques, DAS, data analysis methodology in a manner enabling easy comprehension. Therefore, detailed mathematical treatment and equations have not been discussed.

The endeavour of the monograph is to consolidate information on various sensors, related instrumentation and measurement relevant to chemical lasers application at one place. Instrumentation is vital for the design and implementation of measuring, monitoring and actuation systems and for data acquisition and processing. Rapid advances in computer peripherals have brought about significant changes in the previous Data acquisition and control techniques, which are now much more versatile and user friendly.

The goal of this monograph is therefore to enable scientists and technologists working in rather complex area of chemical lasers to achieve the best technical performances. Till now such topics have been covered scantly in open literature and that too in the research papers only.

Contributors

Mainuddin received his degree in Electronics & Communication Engineering in 1994 from Jamia Millia Islamia, New Delhi and his Masters of Engineering in 2003 from Delhi College of Engineering, Delhi. He received Ph.D. in 2008 from Jamia Millia Islamia, New Delhi. He is presently working as a Professor at department of Electronics and Communication Engineering, Jamia Millia Islamia, New Delhi. He worked as a Senior Research Scientist at Laser Science and Technology Centre, New Delhi for nearly 17 years. His research interests include optical diagnostics, high power lasers, data communication, optical communication and computer networks etc. He has more than 50 publications in international journals/ conferences.

Gaurav Singhal did his graduation in Mechanical Engineering in 1998 from Jamia Millia Islamia, New Delhi and received Ph.D. in 2008 from Indian Institute of Technology, New Delhi. He did his post-doc research work at University of Texas at Austin during 2011-12. He is a Senior Research Scientist at Laser Science and Technology Centre, New Delhi. His research interests include high power lasers, high speed unsteady flows, turbulent mixing, laser diagnostics, CFD techniques etc. He has more than 50 publications in international journals/ conferences.

A. L. Dawar had his post graduation in Physics in 1967 from Kurukshetra University and obtained his Ph.D. in 1974 from Delhi University. He did his post-doc research work at University of Newcastle upon Tyne and University of Glasgow during 1974-75 and 1983-84 respectively. He served as a Senior Scientist at Laser Science and Technology Center, New Delhi. His research interests included high power lasers, thin film, laser materials, integrated optics, EOCM, etc. He has more than 100 publications in international journals & a book on "Semiconducting Transparent Thin Films" to his credit. He retired as senior Scientist from Laser Science and Technology Center, New Delhi in 2004.

Chapter 1

Overview of Chemical Lasers: Sensors and Measurement Needs

Historical Perspective

LASER (Light Amplification by Stimulated Emission of Radiation) is one of the greatest inventions of the twentieth century, which has a variety of applications in human life. During the last few decades, lasers [1] have emerged as the most innovative tool, having wide ranging applications starting from the very common supermarket bar code readers to the highly advanced systems such as nuclear fusion systems for power generation, directed energy weapon systems in an antimissile role. There is hardly any sphere of our life, which is untouched by lasers. A common man encounters lasers not only in light shows, but also with the beauticians, eye specialists, orthopedists, to sight a few examples. Hence, lasers have become an important tool in various fields ranging from a supermarket to star war programs.

Initial work of Charles Townes related to MASER (Microwave Amplification by the Stimulated Emission of Radiation) served as the foundation for Laser Technology. Townes and Schawlow extended the maser concept to optical frequencies in 1958 for which they received the Nobel Prize. Subsequently, in 1960, Theodore Maiman of Hughes Research Laboratory invented the first Ruby laser. There have been a variety of lasers ranging from solid state, semiconductor to liquid and gas lasers. Among these, gas lasers [2-4] offer high flexibility, advantages in cost, good beam quality, and power scalability. He-Ne laser was the first gas laser invented by A. Javan, W. Bennett and D. Harriott of Bell Laboratories in 1961. C. K. N. Patel of the same laboratory invented another gas laser CO_2 laser in 1963. In 1965, Kasper and Pimentel demonstrated the first chemical laser by initiating a hydrogen-chlorine explosion with a flash lamp. Within a few years, HF laser had been demonstrated and by 1984 HF laser with powers greater than 1 MW were developed. The discovery of Chemical Oxygen Iodine Laser (COIL) in 1978 by W. E. McDermott [5] and co-workers resulted in the development of potentially the most powerful chemical gas laser.

The development of lasers in general and high power chemical gas lasers in particular has been a turning point in the history of science and engineering. It has produced a completely new type of systems with potentials for applications in a wide variety of fields. During sixties, lot of work was carried out on the basic development of almost all the major high power lasers including gas dynamic and chemical lasers. Almost all the practical applications of these lasers in defense as well as in industry were envisaged during this period. The motivation of using the high power lasers in strategic scenario was a great driving force for the rapid development of these high power lasers. In early seventies, megawatt class carbon dioxide gas dynamic laser was successfully developed and tested against typical military targets. The development of chemical lasers took slightly longer time because of the involvement of multidisciplinary approach. Though Kasper and Pimental [6] reported the first chemical laser based on flash photolysis of iodine, the major research and technical development, however, took place in seventies only. During a period of ten years from 1970 – 80, chemical lasers particularly hydrogen fluoride, deuterium fluoride matured to an extent that megawatt class system such as Middle Infrared Advanced Chemical Laser (MIRACL) were developed for various military applications.

On the industrial front, lasers are being employed for various material processing applications [7-9] like drilling, cutting, welding, surface hardening, alloying cladding, heat treatment etc. A detailed discussion on all these processes and review of various lasers for this kind of applications are already available in literature [10]. Commercial high power industrial lasers must have qualities such as low running cost, high average power, better beam quality, simple beam delivery system, longer run time and ruggedness. CO_2 and Nd: YAG lasers are the most commonly used in the industry since early sixties for material processing such as cutting, welding, cladding etc. [11]. In the present day scenario, excimer, HF/DF and COIL [9, 12-13] are also being considered for such applications. The analysis carried out by Bohn [14] indicates that CO_2 laser apparently looks most attractive from the cost point of view, however it is worth noticing that the COIL wavelength being much shorter, therefore, its interaction with the materials is much stronger as compared to that of CO_2. Further, COIL wavelength is fiber compatible [15] and hence suitable for remote applications like underwater cutting and dismantling of obsolete nuclear reactor. The running cost (fuel cost) of this laser is high, but it is believed that the regeneration/recycling concepts of fuel would bring down the running cost considerably. Even though the high power Nd:YAG lasers are fiber compatible and have already proven capability for remote operation, the average power of these lasers is in the range of 5 kW and its poor beam quality make them inferior as compared to COIL. COIL has an excellent beam quality which results in better brightness as compared to that of the other potential industrial lasers such as CO_2 or Nd:YAG lasers.

During the last four decades gas lasers have also played an important role in the area of defense through a wide range of equipments starting from Laser Range Finders (LRF) to Directed Energy Weapons (DEW) based on various lasers of different wavelength and power levels. The High-Energy Lasers (HEL) including COIL are mainly aimed as the source for Directed Energy Weapon (DEW) and address applications such as; strategic and tactical missile defense, anti-satellite capabilities, air and ship defense, air craft protection, ground combat support etc.

The major emphasis has been on the development of shorter wavelength CW or high average power pulsed lasers mainly for defense applications. However, with the development of these lasers, the area of material processing has immensely benefited. The most established industrial applications are the laser welding and laser machining including cutting, drilling and shaping. Both for defense as well as material processing applications, it is advantageous to use shorter wavelength with excellent beam quality in order to have high energy density at the desired place.

Chemical lasers are based on chemically generated gases, which are mixed and allowed to flow through the gain region in a direction transverse to the laser beam extraction. Special nozzles are designed to achieve optimal flow conditions of the medium to maximize the efficiency of these lasers. The flow conditions are subject to the kind of chemical laser. For example, in COIL typical cavity conditions are 3 torr pressure with Mach number of 2, whereas, for HF/DF, these typical values are 5 torr and 5 respectively. Toxic and combustible chemicals and gases are generated in such lasers, which must be precisely measured and are required to be either neutralized or stored for later disposal.

Chemically excited iodine lasers are the second-generation lasers in this category. COIL is the first of its kind with electronic transition and was first demonstrated by McDermott [5] in 1978 at Air Force Research Laboratory (AFRL), USA. It is worth mentioning that the lasing from iodine atoms was initially demonstrated, way back in 1964, by Kasper and Pimentel [6] employing UV photolysis of certain alkyl iodide compounds. Later on in 1966, Demaria and Ultee [16] were able to scale up this photolytic iodine laser with flash lamps and produce pulse energy of about 65 Joules. Even though these photolytic iodine lasers have the potential of yielding high-energy laser pulses, the CW operation is limited due to the non-availability of UV sources for continuous pumping. The quest for high power in CW mode had motivated the research in alternative pumping techniques that resulted in the invention of the chemical O_2 (Δ^1_g), as the pumping source for Chemical Oxygen Iodine Lasers (COIL).

Due to their enormous potential, lasers are still among the fast growing areas of present day applied research. Depending on the state of the lasing medium, lasers are mainly classified into solid state, liquid and gas lasers. Among these gas lasers [2-4], Hydrogen Fluoride (HF), Deuterium Fluoride (DF), and Chemical Oxygen Iodine Lasers (COIL) and CO_2 GDL are the only ones to have shown high power capability.

The technological complexity of gas laser development increases with increase in the laser power level. Gas lasers offer the flexibility of operating the lasers in a continuous wave mode because in such lasers heating of the lasing medium can be minimized as compared to that in solid-state lasers and a stable output performance can be achieved for relatively large duration. The electrical efficiency conversion in most solid state lasers is limited to 10 - 15 % and hence the electrical power requirements in case of large output, kind of those needed for defense applications, would require very high capacity electrical generators. But defense applications require these lasers to be taken to the fields and hence it is preferable to minimize the electrical requirements. These requirements were the driving force for the development of chemical lasers. During eighties, chemical lasers particularly hydrogen fluoride, deuterium fluoride matured to an extent that megawatt class system such as Middle Infrared Advanced Chemical Laser (MIRACL) [17] was developed for various military applications. The use of hazardous gas like fluorine in these HF/DF lasers made the researchers to search for an alternate like COIL where the shorter wavelength (1.315 µm as compared to that of 2.7 µm in HF or 3.7 µm in DF) is an additional advantage. This is because the shorter wavelength has better laser material interaction and enables target destruction at relatively low power levels. Further, the development of COIL has accelerated the research in the area of material processing since this laser beam can also be transmitted through optical fibers, which is the limitation in case of HF/DF lasers.

1.1 Hydrogen Fluoride/ Deuterium Fluoride Laser

The goal of all chemical lasers is the efficient conversion of chemical energy into laser energy. High power laser realization necessitates employing chemical reactions, which can produce large energy. The reaction of fluorine with hydrogen has such characteristics (~32 kcal/mole) and it is also quite easy to obtain vibrational inversion of HF molecule for initiating lasing action. This forms the basis for HF and DF lasers [4], produced by mixing hydrogen/ deuterium with fluorine gas. In these lasers, the population inversion is by the chemical reaction, where reaction of atomic fluorine with molecular hydrogen produces vibrationally excited hydrogen fluoride as shown by the equation (1.1). This reaction produces sufficient energy which can be utilized in vibration of newly formed HF bond. Since barrier to this reaction is small, 5 kcal, thus, the rate of producing HF is quite rapid. The HF chemical laser operates at the infrared wavelength of 2.7 µm. The energy level diagram for HF laser is shown in Fig. 1.1.

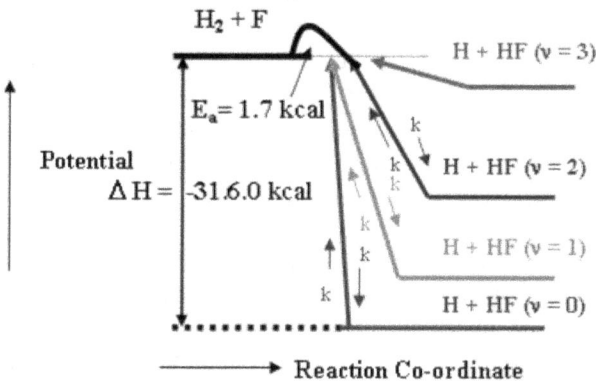

Fig. 1.1. Energy level diagram of HF laser.

The basic Chemical reaction is given by following:

$$F + H_2 \rightarrow HF^* + H \quad \Delta H = \sim -32 \text{kcal/ mole} \quad (1.1a)$$

$$HF^* \rightarrow HF + H + \text{Laser} \quad (1.1b)$$

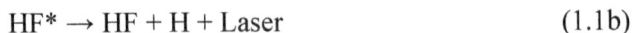

The F atoms must be produced to drive the reaction (1.1a) by thermal dissociation of molecular fluorine at high pressure and temperature. These temperature and pressure conditions are far too high for efficient laser operation, and thus effluents including atomic fluorine are supersonically expanded through a nozzle bank at a Mach 3-5 producing a temperature of 300-500 K and pressure of 3-5 torr in the gain medium. While the flow is cooled, F atoms remain dissociated, as there is insufficient time for recombination to molecular fluorine. Molecular hydrogen is then injected into this supersonic expansion through a large number of very small nozzles to enable good mixing and efficient reaction with atomic fluorine to produce the vibrationally inverted HF molecules for lasing. Another reason for supersonic flow is to stretch the reaction zone over the full width of the laser resonator. Finally, the diffuser is designed to recover the inlet stagnation pressure before the gases are exhausted to vacuum pumps and dumps.

The efficient conversion of the reaction exothermicity into laser radiation requires the judicious control of population inversion, maintaining low temperature of active medium and ensuring uniformity of active medium with sufficient photon flux density. The simultaneous control of all parameters is difficult to achieve. In the flowing medium of a continuous laser, the gas dynamical techniques can be used to optimize the production of excited states, to cool the gas and finally to remove the expended molecules and the waste heat from the cavity. This is the reason that efficient extraction of high power from chemical reactions was first obtained with continuous, gas dynamically controlled HF lasers. Gas dynamical techniques also make it possible to initiate the HF reaction by thermal dissociation of fluorine that resulted in chemical laser of very high power.

A supersonic free jet appears to offer the desired conditions because it combines a fast flow with the possibility of expansion necessary for constant pressure operation. The high pressure and high temperature stagnation conditions can be used to produce the necessary F atoms by thermal dissociation. This thermal dissociation of various fluorine compounds in combination with a fast adiabatic expansion to supersonic speeds has proved the most successful approach. This is because the thermal dissociation under high temperature equilibrium conditions permits the production of large concentrations of F atoms in the most efficient way. The popular thermal drivers include shock tubes, combustors and arc heaters.

However, the arc plasma generator or arc heater is an attractive tool for obtaining the high temperature necessary for dissociation of SF_6 (>2000K). SF_6 is chosen as a fluorine carrier since it is non-toxic, non-corrosive gas that can be handled safely in large quantity. The high temperature, which can be obtained with arc heater, makes it possible to dissociate SF_6 in plenum chamber. To protect the cathode of arc, nitrogen is passed through the arc heater. Moreover, nitrogen gas acts as a buffer gas and dilutes SF_6 to control the plenum temperature. The energy from N_2 arc plasma is utilized to dissociate SF_6 in plenum so that desired numbers of fluorine atoms are generated for initiating the subsequent lasing reaction in the cavity. For a kW level HF arc-driven chemical Laser system, 1 gs^{-1} flow rate of SF_6 is required in plenum to generate desired quantity of fluorine atoms and subsequently 1 gs^{-1} of H_2 is added at the nozzle exit plane to initiate the chemical reaction essential for lasing action. A 50 kW arc heater (input power) is sufficient to create N_2 plasma and to carry out the parametric variations of different gases (viz. SF_6, N_2, O_2 etc) in plenum to have desired lasing composition for kW laser output. Fig. 1.2 shows a schematic block diagram for a kW level HF laser.

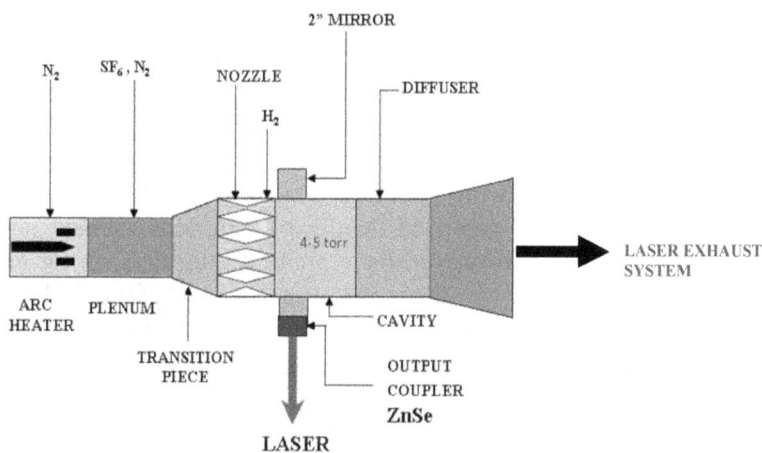

Chemical reaction involved

$H_2 + F \rightarrow HF^* + H$ $\Delta H = -32$ kcal/mole
$HF^* \rightarrow HF + LASER$

Fig. 1.2. Schematic of arc driven HF laser.

DF laser is also based on the same reaction as HF laser but it uses deuterium in place of hydrogen to react with atomic fluorine:

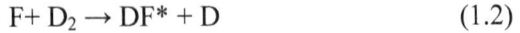

$$F + D_2 \rightarrow DF^* + D \qquad (1.2)$$

This laser may use the similar hardware set up as that of HF laser. The vibrational energy for this laser is less than HF laser and therefore it operates at a wavelength near 3.4 μm. The atmospheric window (3-5 μm) makes this laser suitable for applications requiring atmospheric propagation.

HF/DF lasers, wavelength of 2.6-3.4 μm, lie in the class of high power chemical lasers based on vibrational transition. Their scalability to high power levels enable them to be employed in various defense and industrial scenarios. However, one of their limitations is the toxicity of its constituent elements. Hence, production of lasing species through combustion forms a safety concern. In this context, the use of arc plasma generator is beneficial as it offers safe decomposition of SF_6 for generation of fluorine atoms to be used subsequently for lasing action.

1.2 Chemical Oxygen Iodine Laser

COIL is the first chemical laser of its kind with electronic transition, which was first demonstrated by McDermott [5] in 1978 at Air Force Research Laboratory (AFRL), USA. Lasing is achieved between the electronically excited level of iodine atoms $I(^2P_{1/2})$ and its ground level $I(^2P_{3/2})$ with stimulated emission at 1.315 μm. It has already been stated earlier that, Kasper and Pimentel [6] demonstrated first lasing from iodine atoms by UV photolysis of certain alkyl iodide compounds. Demaria and Ultee [16] extended their work and demonstrated pulse energy up to 65 Joules. However, the search for a mechanism for high power in CW mode led to the invention of the Chemical Oxygen Iodine Lasers (COIL). It uses chemically generated electronically excited oxygen molecules (singlet oxygen), $O_2\,(^1\Delta_g)$, as the pumping source.

This laser uses chemically generated singlet oxygen molecules to pump the lasing medium iodine and hence is classified under the chemical laser group. It involves electronic transition between $I(^2P_{1/2})$ to $I(^2P_{3/2})$ levels. Ogryzlo and his group [18] first reported the reaction of iodine atoms with excited oxygen, produced using electric discharge. They observed a strong emission at 1.315 μm and recognized this as I_2B emission. The excitation of iodine atom is attributed to the near resonant pumping by O_2 ($^1\Delta_g$) molecules. The energy level diagram [18] of the laser is shown in Fig. 1.3.

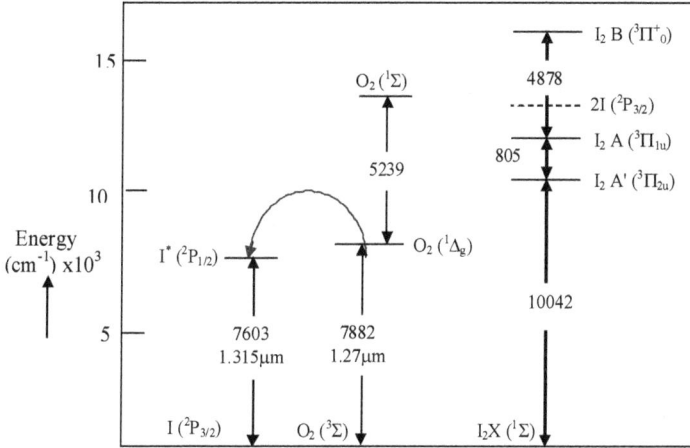

Fig.1.3. Energy level diagram of O_2, I_2 and I in COIL.

This near resonant energy transfer from singlet oxygen molecule to the iodine atoms makes them the most promising pumping source for this laser. In COIL operation, singlet oxygen provides energy both for dissociation of iodine molecules into iodine atoms as well as for excitation of iodine atoms. The following basic reactions are involved in the lasing process:

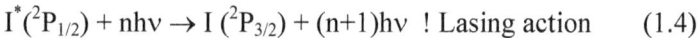

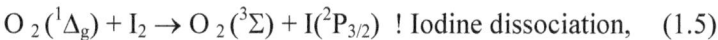

$$O_2(^1\Delta_g) + I(^2P_{3/2}) \Leftrightarrow O_2(^3\Sigma) + I^*(^2P_{1/2}) \quad \text{! Pumping reaction} \quad (1.3)$$

$$I^*(^2P_{1/2}) + nh\nu \rightarrow I(^2P_{3/2}) + (n+1)h\nu \quad \text{! Lasing action} \quad (1.4)$$

$$O_2(^1\Delta_g) + I_2 \rightarrow O_2(^3\Sigma) + I(^2P_{3/2}) \quad \text{! Iodine dissociation,} \quad (1.5)$$

where
$O_2(^1\Delta_g)$ is the singlet oxygen molecule;
$O_2(^3\Sigma)$ is the ground state oxygen molecule;
$I(^2P_{3/2})$ is the ground state iodine atom;
$I^*(^2P_{1/2})$ is the excited state of iodine atom;
I_2 is the iodine molecule.

Some of the important features of COIL that has prompted the rapid development of this laser can be summarized as:

- Higher energy per unit weight (J/kg) as compared to HF/DF laser;
- Comparatively non-toxic;
- Low temperature – low-pressure laser;
- Very compact and light;
- Extremely high chemical efficiency ~ 0.3 [19-21];
- Excellent beam quality;
- The laser output is proportional to the flow of the laser and pumping medium and hence the power scale up is linear;
- Since the laser involves electronic transition, the oscillating wavelength is at near IR region (λ= 1.315 μm), which is fiber compatible;
- Relatively better laser material interaction due to its shorter wavelength;
- More than 200 kW continuous wave power modules have already been demonstrated by AFRL [22]. This potential has led to the use of COIL in various military and civilian applications;
- The possibility of carrying high power beams via optical fibers makes it extremely useful for decommissioning and dismantling of dangerous structures like obsolete nuclear reactors through remote operation [15].

From realization point of view, COIL is a complicated engineering system, which includes many critical subsystems, as shown in Fig. 1.4. The major subsystem includes: the pumping source i.e. singlet oxygen generator, lasing species i.e. iodine supply system, nozzle assembly for the adiabatic expansion of the laser gas, optical resonator / cavity for the power extraction, cold trap for trapping unutilized chlorine & iodine molecules and evacuation system.

The population inversion of the lasing medium (i.e. singlet oxygen and iodine gas mixture) is obtained by the adiabatic expansion of the medium into an optical resonator. COILs in general can operate both in subsonic as well as supersonic regimes. In earlier COILs the lasing medium was flown at a subsonic velocity and therefore termed as subsonic COILs [23-26]. However, subsonic COILs are not only less efficient but also the efficiency is extremely sensitive to the iodine concentration [27-28].

Advanced COILs, on the other hand, are supersonic COILs [29-35] abbreviated as SCOIL, which are much more compact in terms of resonator size and are capable of achieving higher efficiencies. Further, the output power is fairly constant in a wide range of iodine concentration in these systems.

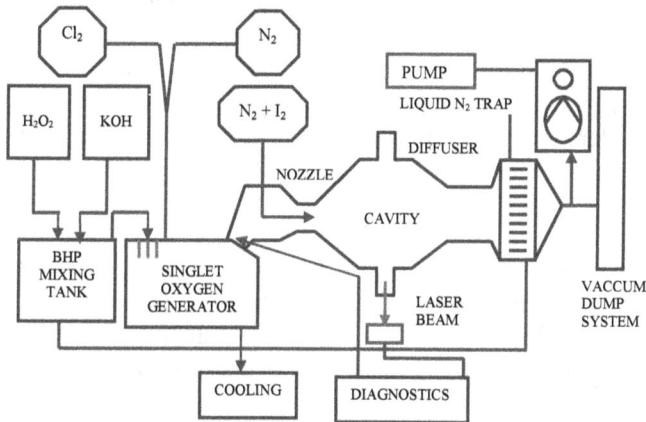

Fig. 1.4. Schematic of COIL system.

In addition, COIL being low pressure and high gas flow system requires high capacity vacuum pumps, which makes the COIL system bulky. In order to achieve compact COIL systems the huge evacuation systems may be replaced by ejector based pressure recovery system, [36-37] which is the core of contemporary COIL research. Pressure recovery system provides the isolation between the low-pressure cavity and laser exit along with ensuring the direct atmospheric discharge of the laser gas flow. US Air force has already developed a COIL laser of ~200 kW power mounted on a Boeing Airbus [38]. The employed ejector based pressure recovery system is comparatively less complicated as compared to a ground-based system since the atmospheric pressure is only about 200 torr. In order to develop ground based mobile systems or industrial COILs the ultimate aim is to achieve an exhaust pressure in excess of one atmosphere (i.e. 760 torr), which is currently a challenging COIL research area.

From the first realization of this laser, there has been significant progress in the development of this laser because of worldwide interest in the shortest wavelength and high efficiency. With the introduction of supersonically flowing (Supersonic COIL) laser gas (i.e., singlet oxygen and iodine gas mixture) there is a large potential for power scaling up. The gain in the cavity is inversely proportional to temperature and thus the adiabatic expansion of the lasing medium helps in higher gain.

Even to this day the most advanced COILs are based on the same principles as suggested by McDermott, except that the techniques for producing singlet oxygen have improved from bubbler generator to jet / spray generator. Efforts are also being made to make the system more efficient as well as alternatives are being tried to dispense with Basic Hydrogen Peroxide and chlorine for the production of singlet oxygen. These include the concept of electri- COIL [39] and All Gas Iodine Laser (AGIL) system approach [40].

The first demonstrated COIL power at US Air Force laboratory was only few mill watts and over a short span of 27 years, modules of 200 kW power have been developed to push the power levels to a megawatt class. COIL technology has undergone numerous improvements and chemical efficiencies as high as 33 % have been already demonstrated [41]. In addition, COIL power was also demonstrated under various modes like Q-switching, mode locking, frequency doubling etc.

Chemical Oxygen Iodine Laser (COIL) is an advanced high power laser source involving cutting edge technologies and advanced research is being carried out worldwide as a part of the program "Directed Energy Weapon". Chemical Oxygen Iodine Laser is the most potential laser among the high power lasers, which is explored not only with the aim of its use in defense but also for its wide applications in civil area.

1.3 CO_2 Gas Dynamic Laser

CO_2 laser is basically one of the gas lasers and it does not fall truly in the category of chemical lasers. However, it utilizes benzene and nitrous oxide as one of the fuel in CO_2 gas dynamic laser [3]. But most of the gas dynamic aspects of this laser are similar to HF-DF laser and from sensor and measurement point of view there is a similarity among CO_2 GDL, HF-DF and COIL lasers. It is the reason why CO_2 GDL is also being dealt with in this monograph.

This laser is commercially popular because of its high efficiency (5-20 %) and high power outputs (1 W to 100's of kW). In addition to high efficiency, CO_2 laser enjoys the benefit of removal of waste heat generated during laser excitation. The flowing gas through the laser can remove waste heat efficiently; in fact, this is the common advantage of all the gas lasers.

CO_2 laser operates in mid-infrared regions and is based on rotational-vibrational transitions at wavelength between 9 μm and 11 μm, however, the strongest emission is at 10.6 μm. These wavelengths are strongly absorbed by organic materials, water and organic tissue and hence it can be used for application such as cutting plastics or performing medical surgery. The lasing wavelength of CO_2 laser falls in a band where atmospheric attenuation is very little. Hence, this laser also finds applications in wireless communication system such as optical radar. High power CO_2 lasers are also used in industries for welding, drilling, cutting etc.

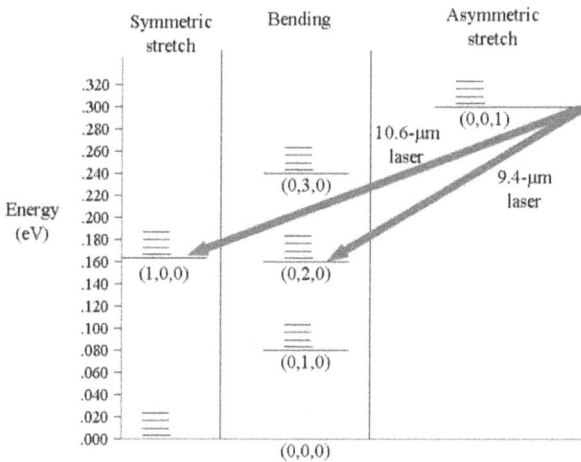

Fig. 1.5. Energy level diagram of CO_2 molecule.

CO_2 molecule has three mode of vibration: (i) Symmetric stretching mode (ii) Bending mode and (iii) Asymmetric stretching mode. Each vibrational mode has its own quantized energy levels as shown in Fig. 1.5. It shows the energy level diagram of CO_2 molecule. If the excited molecule drops to the first excited state of symmetric stretching mode, it results in release of 10.6 μm laser radiation. If it drops to the second excited level of bending mode, a laser radiation of wavelength of 9.4 μm is emitted. However, the 10.6 μm emission is stronger than other transitions.

The population inversion is achieved by a number of methods: it may be dc excited, RF excited or by gas dynamic means. In case of electrical excitation, addition of nitrogen molecule increases the efficiency of this laser because lowest vibrational level of nitrogen matches with the energy required for exciting the CO_2 molecule to upper laser level. However, typical voltages are kilovolts or more and typical gas pressure is kept low to sustain a stable continuous discharge. Helium is also added in gas mixture due to its better thermal conductivity and thereby resulting in efficient heat removal. Moreover it helps in depopulating the lower laser level, thereby increasing the population inversion.

An alternate method for generation of very high laser power (100's of kW) is by thermal expansion of gases. Basically, gases at very high temperature (1700-1800 K) and high pressure (25-35 bar) are passed through a supersonic nozzle and there is sudden expansion of mixture, which results in very low temperature (~300 K) and pressure (30-40 torr). This sudden change results in population inversion of CO_2 and results in laser power at 10.6 μm. This type of CO_2 laser is called CO_2 gas dynamic laser. Fig. 1.6 and Fig. 1.7 shows basic principle and schematic of a combustion based CO_2 GDL system respectively. In this laser, a mixture of Benzene (C_6H_6), Nitrous oxide (N_2O) and Nitrogen (N_2) is passed through a combustor at ~1700 K and 35 bar. This results in production of CO_2, N_2 and water (H_2O). The active medium comprising of these combustion products is passed through a supersonic nozzle bank at a Mach number ~ 4-5 producing required number density of CO_2 molecule in the cavity for lasing. A diffuser is employed at the exit of the laser cavity for direct exhaust of this gas mixture to the atmosphere (760 torr).

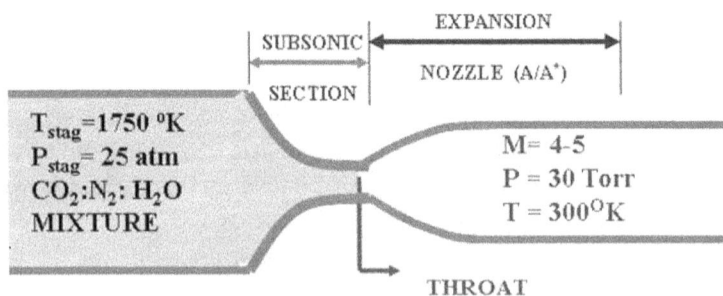

Fig.1.6. Basic concept of CO_2 GDL.

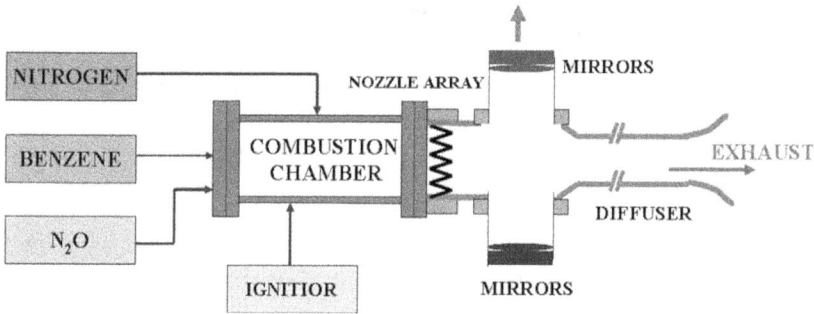

Fig.1.7. Schematic diagram of CO_2 GDL.

Table 1.1 shows the comparison of CO_2 GDL, HF-DF and COIL laser. It is clear that COIL has maximum efficiency as far as the energy/mass ratio is concerned. However, it imposes requirement of large pumps/dumps or active pressure recovery system for direct exhaust from cavity pressure of 3 torr to atmosphere. Although, CO_2 GDL is non toxic but its size is very bulky for very high laser power. HF-DF lasers are not preferred because it utilizes highly toxic chemicals and gases. Till date COIL seems to be a better choice for high power applications.

Table 1.1. Comparison of COIL, HF-DF Laser, CO_2 GDL.

Laser	Typical Wavelength (in μm)	Efficiency (kJ/kg)	Safety Hazards	Operational complexity
CO_2 GDL	10.6	30-40	Non-Toxic	Proven Technology, but High Temperature and High pressure requirements
HF-DF laser	3.4-DF 2.7-HF	300-DF 200-HF	Highly Toxic Highly Toxic	Complex Technology & Hazardous
COIL	1.315	490	Low Toxic	Complex emerging technology, but requirement of large exhaust discharge

1.4 Sensors and Measurement Needs

As discussed in previous sections, the flowing medium gas lasers employing various gas effluents and chemical reagents offer high flexibility, advantages in cost, better beam quality, and power scalability.

The efficient conversion of the reaction exothermicity (energy) into laser radiation requires the judicious control of following contributing factors:

1) The production rate should be fast compared to deactivation;

2) The temperature of active medium has to be kept low;

3) The photon flux density in the cavity should be high;

4) The active medium should be uniform.

The simultaneous control [42-44] of all these parameters is difficult to achieve. In the flowing medium of a continuous laser, gas dynamic methods can be used to optimize the production of excited states, to cool the gas and finally to remove the expended molecules and the waste heat from the cavity. This allowed efficient extraction of cw high power from chemical reactions with gas dynamically controlled gas lasers. Gas dynamical techniques also make it possible not only to create the desired lasing conditions but also enable developing lasers of very high power. But the criticality of gas flow chemical lasers lies in:

a) Producing Pumping medium (e.g. singlet oxygen molecule in COIL);

b) Producing suitable laser medium (e.g. iodine in COIL);

c) Maintaining required concentration of lasing medium;

d) Maintaining required concentration of pumping medium;

e) Maintaining the pressure uniformity in a supersonic flow;

f) Converting a sizeable fraction of the exothermicity of reaction into laser power;

g) Control explosive nature of its constituent chemicals;

30

h) Implementing safety schemes as hazardous chemical, gases are involved.

Thus the flow parameters of various gas effluents viz. flow rate of individual gas constituents, pressure, temperature and Mach number are very crucial in determining the output of these high power infrared gas lasers. The basic parameter measurement such as pressure may provide critical insight into complex phenomenon such as shock waves typically encountered towards the exit section in a chemical laser flow field. A series of pressure probes placed along say an ejector channel provides information about the location at which the shock wave is anchored, which may be of great aid towards system optimization by altering the input parameters. Similarly, the pressure, temperature and flow rates for pumping and lasing species for a given configuration may be coupled to estimate the penetration of lasing species into the pumping species and hence the mixing efficiency of the two streams. The gas stream conditions may then be suitably altered to maximize the laser output.

Also, there is a need for on line estimation of specie concentration of the lasing and pumping mediums depending on the laser, using optical/non-optical methods. For example, in case of COIL, an optical absorption spectroscopy (at wavelength 499 nm) based diagnostics [45] may be used for on line estimation of iodine (lasing medium). The laser power in case of COIL is highly dependent on the optimal iodine concentration since both excess and lack of iodine degrades laser output. The online monitoring of iodine enables the maintenance of ideal iodine flow rates, which may vary according to system design considerations.

Similarly, pumping specie for this laser i.e. singlet oxygen emits at 1.27 μm, for which emission spectroscopic based sensors can be applied. Since all of these lasers discussed in previous sessions involve intensive reactive processes which may be multiphase as well, hence determination of concentration of chemical species is not limited to lasing /pumping but may also involve intermediate reactants/ products, water vapor etc. One such example is measurement of chlorine utilization in COIL flows at exit of the singlet oxygen generator using absorption spectroscopy at 330 nm. The measurements of both singlet oxygen yield and chlorine concentration together define the efficiency of the singlet oxygen generator and determine the input power available for lasing.

Apart from basic parameter sensing and measurement, these advanced laser systems also require customized techniques for evaluating the laser performance. This is essential since many of the questions regarding their operation are still unanswered requiring sensitive diagnostics, which may provide critical insights into these scientifically interesting laser systems. A typical example of such diagnostics is laser-induced fluorescence (LIF), planar laser induced fluorescence (PLIF) of various interacting gas streams. Recently, laser based non-intrusive diagnostic techniques such as these are becoming widely prevalent in order to gain insight into visibly complex flow fields associated with these chemical lasers. This may enable assessing the occurrence of various scales of turbulence, understanding mixing of various interacting gas streams and degree of uniformity achieved, occurrence of shock waves and their interaction with mixing layers, investigation of flow separation especially in the regions of adverse pressure gradients.

Also, a similar diagnostics may be employed for determining the small signal gain, a parameter central to the operation of all these lasers. An important role of gain diagnostics occurs in optimizing the location of the optic axis for laser power extraction. This is done by measuring the gain at various locations along the direction of the flow. It may easily affect the laser power by 10-20 % and occasionally even more.

Other such diagnostics may include power measurement for such high power cases, water vapor concentration just to name a few.

The real time monitoring, acquisition, estimation and display of parameters, such as gas specie concentration, Mach number, mass/molar flow rate, pressure, temperature etc., is also central for desired functioning of the laser. This function is performed by the data acquisition system (DAS), which works in complete sync with all the sensors, placed at various locations across such complex systems and displays all the vital system parameters. It acquires all the operating parameters of a chemical laser system through various individual sensors and converting the data received thereof into the desired form and displaying it to the user for easy interpretation. Consider, the measurement of iodine concentration, which involves determining the pressure, temperature and buffer gas flow rate using individual sensors. The iodine molar flow is subsequently calculated by the data acquisition system utilizing all the information from on site sensors for pressure, temperature and flows and displayed as temporal plot for easy comprehension. Further, it also enables making on line adjustments in flow rates by utilizing requisite feedback mechanism. Moreover, as illustrated above with an example, it represents the acquired data in graphical form with high temporal resolution for analyzing and understanding the performance of these lasers.

It is also necessary to emphasize that the typical duration of operation of these lasers is short (few seconds in one shot) due to various reasons. Hence, DAS also provides a method for sequential and switching control of various fuel supplies, which is very critical for optimum performance of the system.

References

[1]   Orazio Svelto, Principle of Lasers, *4th Edition, Plenum Press*, New York, 1998.

[2]   David D. Skatrud and Frank C. De Lucia, Excitation, Inversion, and Relaxation Mechanisms of the HCN FIR Discharge Laser, *Appli. Phys. B*, 35, 1984, pp. 179.

[3]   Garry E. T., Gas Dynamic Laser, *Journal of IEEE Spectrum*, 7, 1970, pp. 51.

[4]   Wilson L. E and Hook D. L., Deuterium Fluoride CW Chemical Lasers, *AIAA Conference*, Santiago, 1976, paper 76-344.

[5]   McDermott W. E., Pchelkin N. R., Benard D. J., Bonsek R. R., An electronic transition Chemical Laser, *Applied Physics Letters*, 32, 8, 1978, pp. 469.

[6]   Kasper, J. V. V., Pimental, G. C., *App. Phys. Let.*, 5, 1964, pp. 231- 233.

[7]   Kar A., Scott J. E. and Latham W. P., Theoretical and Experimental Studies of Thick-Section Cutting with a Chemical Oxygen-Iodine Laser (COIL), *Journal of Laser Applications*, Vol. 8, No. 3, June, 1996, pp. 125.

[8]   Carroll D. L. and Rothenflue J. A., Experimental Study of cutting thick aluminium and steel with a Chemical-Oxygen-Iodine Laser using N_2 or O_2 Gas Assist, *Journal of Laser Applications*, Vol. 9, No. 3, 1997, pp. 119.

[9]   Yasuda K., Atsuta T., Sakurai T., Okado H., Hayakawa A. and Adachi J., Study on material processing of Chemical Iodine Laser, in *Proceedings of the 3rd JSME/ASME Joint International Conference on Nuclear Engineering*, 1995, pp. 1769.

[10]   Premasundaran M., High power lasers and their applications, 1st Edition, *Law and Commercial Publications*, New Delhi, India, 2004, pp. 105.

[11]   Belforte D., Levitt M., The Industrial Laser Annual Handbook, *Penn Well Books*, Tulsa, Oklahoma, 1990.

[12]   Joeckle R., Gautier B., Nett J., Schellhorn M., Sontag A., Stern G., Laser material interaction with short wavelength infrared laser, *AIAA paper*, 1995, pp. 95.

[13]   Vetrovec J., Hindy R., Subbraman G., Spiegel L., High power Iodine laser application for remote D & D cutting, *SPIE*, 3092, 1996, pp. 780.

[14]   Bohn W. L., German COIL efforts: Status and Perspectives, *SPIE*, 4631, 2002, pp. 53-59.

[15]   Endo M., Tei K., Sugimoto D., Nanri K., Uchiyama T. and Fujioka T., Development of a prototype COIL for decommissioning and dismantlement, *SPIE*, 4184, 2001, pp. 23.

[16]   Demaria A. J. and Ultee C. J., High energy atomic iodine photo dissociation laser, *Applied Physics Letters*, 9, 1966, pp. 67.

[17]   http://www.abovetopsecret.com/pages/miracl.html

[18]   Arnold, S. J., Finlayson N. and Orgryzlo E. A., Some Novel Energy Pooling process Involving $O_2(^1\Delta_g)$, *Journal of Chemical Physics,* 44, 1966, pp. 2529.

[19]   Fujji H., Yoshida, S., Iizuka, M., Atsuta, T., Development of high power chemical oxygen iodine laser for industrial application, *Journal of App. Phys.,* 67, 1990, pp. 3948.

[20]   D. K. Monroe, Space debris removal using a high power ground based laser, *SPIE,* 2121, 1994, p. 276.

[21]   Tei K., Sugimoto D., Endo M., Takeda S. and Fujikova T., Chemical Oxygen Iodine Laser for decommissioning and dismantlement of nuclear facilities, *SPIE*, 3887, 1999, pp. 162.

[22]   McDermott, W. E., Stephens, J. C., Vetrovec, J., Dickerson, R. A., Operating experience with a high throughput jet generator, *SPIE,* 3092, 1997, pp. 146.

[23]   Benard, D. J., McDermott, W. E., Pchelkin, N. R., Bousek, R. R., Efficient Operation of a 100W Transverse Flow Oxygen Iodine Chemical Laser, *Applied Physics Letters*, 34, 1979, p. 40.

[24]   Rosenwaks S. and Bacher J., An efficient small scale Chemical Cxygen Iodine Laser, *Applied Physics Letters*, 26, 1982, pp. 87.

[25]   Bonnet J., Experimental analysis of a Chemical Oxygen Iodine Laser, *Applied Physics Letters*, 46, 1984, pp. 1009.

[26]   Schmiedberger. J., Kodymova, J., Spalek, O., Kovar, Experimental study of gain and output coupling characteristics of a CW Chemical Oxygen Iodine Laser, *IEEE Journal of Quantum Electronics*, 27, 6, 1991, pp. 1265.

[27]   Zagidullin M. V., Nikolaev V. D., State of the art and perspectives of Chemical Oxygen Iodine Lasers, *SPIE*, 3688, 1999, pp. 54.

[28]   Zagidullin M. V., Nikolaev V. D., Svistun M. I., Khvatov N. A., Comparative characteristics of subsonic and supersonic oxygen-iodine lasers, *Quantum Electronics*, 28, 5, 1998, pp. 400.

[29]   Elior A., Lebiush E., Schall W. O., Rosenwaks S., A small scale, supersonic Chemical Oxygen Iodine Laser, *Optics and Laser Technology*, 26 1994, pp. 87.

[30] Balyvas, I., Barmashenko, B. D., Furman, D., Rosenwaks, S., Zagidullin, Z., Power optimization of small scale Chemical Oxygen Iodine Laser with jet type singlet oxygen generator, *IEEE Journal of Quantum Electronics*, 32, 1996, pp. 2051.

[31] Rittenhouse T. L., Phipps S. P., Helms C. A., and Trusdell K. A., High Efficiency operation of a 5 cm gain length supersonic Chemical Oxygen Iodine Laser, *SPIE*, 2702, 1996, pp. 333.

[32] Lebiush E., Barmashenko B. D., Elior A., Rosenwaks S., Parametric study of the gain in a small scale, grid nozzle supersonic Chemical Oxygen Iodine Laser, *IEEE Journal of Quantum Electronics*, 31, 1995, pp. 903.

[33] Gaurav Singhal, Mainuddin, R. Rajesh, A. K. Varshney, R. K. Dohare, Sanjeev Kumar, V. K. Singh, Ashwani Kumar, Avinash C. Verma, B. S. Arora, M. K. Chaturvedi, James Sudhir, R. K. Tyagi, A. L. Dawar, Test bed for a high throughput supersonic chemical oxygen-iodine laser, *Quantum Electronics*, 41, 5, 2011, pp. 430.

[34] R. Rajesh, Gaurav Singhal, Mainuddin, R. K. Tyagi, A. L. Dawar, High throughput jet singlet oxygen generator for multi kilowatt SCOIL, *Journal of Optics and Laser Technology*, Vol. 42, 2010, pp. 580.

[35] R. K. Tyagi, R. Rajesh, Gaurav Singhal, Mainuddin, A. L. Dawar and M. Endo, Supersonic COIL with Angular Jet Singlet Oxygen Generator, *Journal of Optics and Laser Technology*, Vol. 35, 5, 2003, pp. 395.

[36] Yang T. T., Dickerson R. A., Moon L. F. and Hsia Y. C., High Mach number, High Pressure Recovery COIL Nozzle Aerodynamic Experiments, *AIAA 30th Plasma dynamics and Lasers Conference*, 2000, pp. 2000.

[37] Gaurav Singhal, Mainuddin, R. K. Tyagi, A. L. Dawar, P. M. V. Subbaroa, Pressure recovery studies on a supersonic COIL with central ejector configuration, *Journal of Optics and Laser Technology*, Vol. 42, 2010, pp. 1145.

[38] Stephen C. Hurlock, COIL technology development at Boeing, *SPIE*, 4631, 2002, pp. 101-115.

[39] Carroll D. L. and Solomon W. C., ElectriCOIL: An advanced chemical iodine laser concept, *SPIE*, 4184, 2001, pp. 40.

[40] Hanshaw T. L., Madden T. J., Herebelin J. M., Manke G. C., Anderson B. T., Tate R. T., Hager G. D., Measurement of gain on the 1. 315 μm transition of atomic iodine as produced from NCl (^{1}aΔ) + I(^{2}P$_{3/2}$) energy transition, in *Proceedings of the SPIE*, 3612, 1999, pp. 147.

[41] Endo, M., Osaka, T., Takeda, S., High efficiency Chemical Oxygen Iodine Laser using stream wise vortex generator, *Applied Physics Letters*, 84, 16, 2004, pp. 2983.

[42] Mainuddin, Gaurav Singhal, R K Tyagi, Data acquisition system for flowing gas lasers, in *Proceedings of the 3rd IEEE International Conference on Electronics Computer Technology (ICECT' 2011)*, Kanyakumari, Vol. 2, 2011, pp. 184-188.

[43]   Mainuddin, Gaurav Singhal, R. K. Tyagi, and A. K. Maini, Diagnostics and data acquisition for chemical oxygen iodine laser, *IEEE Transactions on Instrumentation and Measurement*, 61, 6, 2012, pp. 1747.

[44]   Mainuddin, R. K. Tyagi, R. Rajesh, Gaurav Singhal and A. L. Dawar, Real-time data acquisition and control system for a chemical oxygen-iodine laser, *Measurement Science and Technology*, 14, 2003, pp. 1364.

[45]   Mainuddin, M. T. Beg, Moinuddin, R. K. Tyagi, R. Rajesh, Gaurav Singhal and A. L. Dawar, Optical spectroscopic based In-line iodine flow measurement system-an application to COIL, *Sensors and Actuators*: B, Vol. 109, 2005, 2005, pp. 375.

Chapter 2

Direct Sensors: Types and Selection

Advances in science and technology are deeply interwined. Sensor technology along with computer processing allows us to visualize physical phenomenon we are unable to otherwise comprehend.

Contemporary laboratory research and development in the state of the art technologies like chemical lasers, manufacturing process control, analytical instrumentation and aerospace technology all would have diminished capability without the availability of suitable modern sensors and measurement techniques.

It is important to understand that all the sensors and transducers along with data acquisition form an instrument to control and optimize scientific or technological devices. Hence, the sensors/transducers and data acquisition are a crucial instrument in development of complex chemical gas laser systems. In chemical gas lasers, the temperature, pressure, flow rates of individual gas constituents are very critical in determining the output of these high power infrared gas lasers. The optical signal detection is also essential for development of non-contact type measurement techniques.

Hence suitable sensors are required to perform testing and optimization of chemical gas lasers. The prime aims of measurements in chemical lasers may be outlined as under:

1) Confirming theoretical models;

2) Evaluation of degree of correctness of design;

3) Evaluation of performance of chemical laser;

4) Optimization of system parameters based on measured values;

5) Enhanced understanding of complex phenomenon associated with laser operation;

6) Revealing unanswered facts behind critical technological grey areas

7) Collecting exhaustive data for new and improved design.

In order to fulfill these aims, measurement should be carried out carefully with latest available techniques. Accurate measurement and subsequent control actions are ladders of success of an engineering design of chemical gas lasers.

This chapter deals with few of the most fundamental concepts pertaining to electrical sensors or transducers and also discusses various types of direct sensors viz., temperature, pressure, level, flow and optical sensors. Their respective classification and principle of operation have also been dwelled upon. Further, sensor selection is another important task which is carried out on the basis of the application requirements. The measurement range, accuracy, response time, compatibility with the measured medium and cost are the main factors, which decide the type of sensor. Thus, the present chapter also deals with the selection of various sensors pertaining to the needs of different chemical gas lasers.

2.1 Sensor Fundamentals

Sensors and transducers [1-5] are words, which are quite often used interchangeably. There is no hard and fast rule, which distinguishes an electrical transducer from an electrical sensor. A sensor is a device that detects or measures a physical quantity and an electrical sensor is a device that converts a physical quantity in to an electrical output signal. The output signal from the sensor is determined by the sensor's electromechanical or chemical properties. The variation in physical parameters may directly generate an electrical quantity due to thermoelectricity, piezoelectricity or photoelectricity or it may result in change of intermediate quantity due to variation in intrinsic properties (resistivity, permeability, permittivity or refractive index) of sensing device.

A transducer is a device that can receive one type of energy and convert it in to another type of energy. An electrical transducer is a device, which consists of sensor and signal conditioning so that desired electrical signal can be obtained at output. It is capable of modifying the sensor's output signal by converting that signal into a more desirable form. We can say that a sensor is basically a subset of a transducer. Fig. 2.1 shows a typical transducer configuration. Let us take an example of a thermocouple, which senses temperature and produces an emf, which is directly proportional to the temperature difference between its junctions. This thermocouple is a sensor.

However, since the output of thermocouple (typically μV) is not in the desirable form, one can amplify this output using an instrumentation amplifier, which not only provide amplification but also improve common mode rejection ratio as well as isolation between input and output. This combination of thermocouple with instrumentation amplifier and power supply is basically a transducer unit that suffices the requirement of output voltage signal in desired range.

Transducer

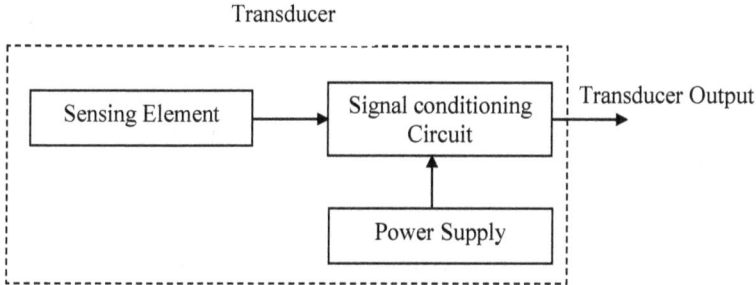

Fig. 2.1. Typical transducer configuration.

2.1.1 General Classification

A number of sensors and transducers are available for use in different applications. There are various ways in which these may be classified. One classification that may be based is on the basis of the physical parameter being measured such as *temperature, pressure, light* and so on. Other classification is based on the type of electrical signal being produced by the sensor or transducer such as *voltage* or *current*. Third classification may be based on whether the sensor is *contacting* or *non-contacting*. A *contact type* sensor is in direct contact with the medium and hence it may not only disturb the medium (if medium is flowing as in case of chemical gas lasers) but also may suffer from mechanical or chemical wear and tear. On the other hand, a *non contact* type sensor is the one which is not in direct physical contact with the medium, an example in case is optical based sensors employed for non intrusive detection, where a contacting sensor is likely to disturb the medium. It is essential to implement *non-contact* type sensors and measurement technique in chemical gas lasers primarily owing to the nature and the physical state of the lasing medium. Non-contact type optical sensors are discussed in detail in Chapter 3. Fourth classification is based on whether the sensor requires excitation or not. Sensors can be categorized into two types, namely *active* and *passive* sensors/transducers. *Active* sensors are those that do not require external excitation to operate for example thermocouple. *Passive* sensors require external excitation for its operation for example resistance temperature detector (RTD) and strain gauges.

It is appropriate to mention that in this monograph the sensors/ transducers which provide a direct electrical output, with minimal signal conditioning, for a change in physical parameter being measured are classified as *direct sensors/ transducers*. All those sensors or transducers, which process the signal inputs from several direct sensors to evaluate a physical parameter (also termed as *derived parameter*) are classified under *diagnostics* (indirect sensors). Therefore, to cite an example, a pressure sensor will be classified as a *direct sensor* where as singlet oxygen concentration measurement unit is classified as a *diagnostic* since it requires inputs from a photo-detector, pressure and temperature sensor to compute singlet oxygen concentration.

2.1.2 Transducer Characteristics

The selection criterion for the type of transducer employed depends upon the physical parameter to be measured, cost and application requirements. The main characteristics that are most important in determining the applicability of the transducer are given below:

Accuracy

Accuracy is an important aspect of measurement of any physical parameter. Accuracy refers to how closely the measured value agrees with the true value of the parameter. It describes the maximum error that can be expected from a measurement taken at any point within the operating range of the transducer. Accuracy is normally defined as a percentage error over the operating range of the transducer such as $\pm$ 1 % between -50 °C to 100 °C of temperature range. Thus, percentage accuracy of a sensor or transducer is defined as a percentage error to the actual value.

Sensitivity

Sensitivity of a transducer is basically its ability to produce an output response for a given input change. Thus, it is expressed as the ratio of change in output signal to the change in input signal of the transducer.

Repeatability

Repeatability of a transducer is defined as the degree of agreement among independent measurements of a quantity under the same condition. The ability of the transducer to generate almost identical output responses to the same physical input throughout its working life is an indication of its reliability and is directly related to the cost of the transducer.

Reproducibility

Reproducibility in a transducer is the closeness of agreement among the measurements of the same quantity at different times in the same environmental conditions.

Range

A transducer is usually designed to operate over a specified range. The range of a transducer refers to the minimum and maximum values of the input variable for which it has been designed and the defined limits of all characteristics are met. For example, a thermocouple can work well beyond its specified operating range of -200 °C to 850 °C, however, its sensitivity may be too small to produce the accurate and repeatable measurements.

The accuracy, sensitivity and repeatability of the measurement are affected by several variables. Transducer itself may disturb the physical quantity being measured as in case of temperature sensor by heating its own mass. Transducers are also affected by unwanted noise. Even excitation signals can result in self-heating of the devices, such as excitation current may lead to heating in case of RTDs.

The following sections discuss various types and selection of temperature sensors, pressure sensors, level sensors, flow sensors and photo-sensors.

2.2 Temperature Sensors

The physical quantity 'Heat' is one of the forms of energy and measured in terms of joules. The change in heat content of any object results in a change in temperature. Hence, temperature of an object is a measure of heat energy content of an object. The temperature and energy are analogous to voltage and electrical energy. The most commonly used sensor amongst all type of sensors is the one detecting temperature (or heat) i.e. temperature sensor.

The electrical parameters of temperature sensors [6] vary in a particular fashion with temperature. Thermocouples, RTDs and thermistors etc. are the most common examples of temperature sensors.

2.2.1 Thermocouples

The most widely used temperature sensor in industry is thermocouple [7]. A thermocouple is a temperature sensor consisting of two dissimilar materials that are in thermal and electrical contact. A potential develops at the interface of two materials as the temperature changes. This thermoelectric phenomenon is known as the thermoelectric effect or *Seeback effect* which was discovered by the German-Estonian physicist Thomas Johann Seebeck in 1821. The amount of voltage that a thermocouple produces depends on the two types of metal that are used to form the junction. The voltage developed across the junction can be measured by the arrangement shown in Fig. 2.2.

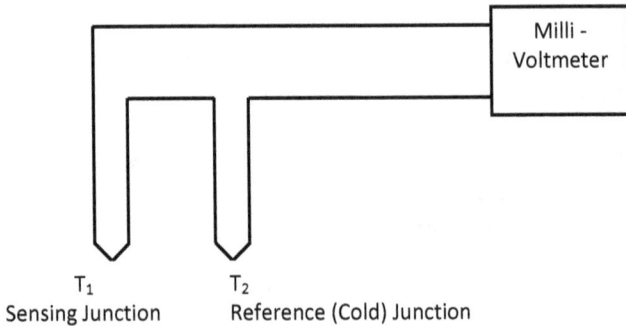

Milli - Voltmeter

T_1
Sensing Junction

T_2
Reference (Cold) Junction

Fig. 2.2. Voltage measurement in thermocouple.

A milli-voltmeter is connected across the junction T_1 and T_2. When two junctions (T_1 and T_2) are in a circuit with different junction temperatures, a voltage of few mV may be detected. This voltage will be zero if junctions are at the same temperature and it will increase as the temperature of one of the junction (T_1) relative to other junction (T_2) is changed. The second junction can be controlled to keep it at a constant temperature (normally 0 °C). Initially, it used to be done by placing the junction into an ice bath, and hence it was popularly known as cold junction. Now-a-days same function is accomplished using electronic compensation circuit (employing thermally sensitive device such as a thermistor), which is known as cold junction compensation circuit. The cold junction compensation circuit produces an equivalent voltage as produced by junction in ice bath. Hence, the voltage from a known cold junction can be simulated, and the appropriate correction applied. Hence, actual output voltage produced is because of the temperature of junction (T_1) only.

Thermocouples are available in a wide range of different types, constructions and temperature ranges. Thermocouples are robust, low cost and have very wide operating range. In contrast to most other methods of temperature measurement, thermocouples are self powered and require no external form of excitation. The main limitation with thermocouples is accuracy, specifically, system errors of less than one degree Celsius can be difficult to achieve.

The various types of thermocouples fall into two categories: base metal types such as iron-constantan (J, K, N, E & T) and noble metal types such as platinum/rhodium-platinum (R, S & B). Noble metal type (e.g. type R) thermocouples are required for the higher temperatures but have lower output levels. But they have the potential for great accuracy and are more often used in laboratories, but not as much in industry due to their much higher cost. In addition, there are several high temperature tungsten-based thermocouples (type G, C and D), which allow temperature measurements between 0 °C and 2800 °C. However, they are not suitable to be employed in an oxidizing atmosphere. Thermocouples have sensitivities from a few μV per °C to a few 10's of μV per °C, varying with type and operating temperature.

Ideally, the thermocouple wires should connect directly to the instrumentation, but in a practical scenario this is not the case and the wires need to be extended. Extension cable is made from the same thermocouple materials. It has better performance but is generally associated with higher costs. Table 2.1 shows a comparison between various commonly used thermocouples, which would enable in making a suitable selection.

Table 2.1. List of various available thermocouples, operating ranges and sensitivities.

Types of Thermocouple	Temperature Range (°C)	Typical Sensitivity	Remarks, if any
Type T (Copper-Constantan) (Cu)-(Cu, 45 % Ni)	(-185 to 400)	46 µV	Very stable & is suitable for low temperature application. Since both conductors are non-magnetic, there is no Curie point and thus no abrupt change in characteristics but emf/temperature curve is quite nonlinear. Useful in food, environment & refrigeration applications.
Type J (Iron-Constantan) (Fe)-(Cu, 45 % Ni)	(-185 to 870)	55 µV	Mostly used for calibration purposes. Suitable for use in vacuum, inert environment. Oxidization with iron and thermal drift is excessive.
Type K (Chromel-Alumel) (Ni,10 %Cr)-(Ni, 2 %Mn, 2 %Al, 1 % Si)	(-185 to 1260)	42 µV	Increased upper temp. limit but reduced sensitivity. Its emf/temperature curve is reasonably linear.
Type E (Chromel-Constantan) (Ni, 10 % Cr)-(Cu, 45 % Ni)	(0 to 980)	67 µV	Largest output voltage but upper temp. limit is ~1000 °C only. Additionally it is non-magnetic and hence advantageous in some applications.
Type N (Nicrosil-Nisil) (Ni, 14 % Cr, 1 % Si)-(Ni, 4 % Si, 0.1 % Mg)	(-270 to 1300)	30 µV	Suitable for accurate measurement in air up to 1200 °C, which is slightly, lower than K-type. Less aging effect than K type, stable for nuclear environment, oxidation resistant, very costly
Type S (Pt 10 % Rhodium -Pt) (Pt, 10 % Rh)-(Pt)	(0 to 1535)	6-12 µV	Easily contaminated, protected with gas tight ceramic tubes. Used as the standard of calibration for the melting point of gold
Type R (Pt 13 % Rhodium-Pt) (Pt, 13 % Rh)-(Pt)	(0 to 1590)	6-14 µV	Higher emf than type S, Easily contaminated, protected with gas tight ceramic tubes.

2.2.2 Resistance Temperature Detectors (RTDs)

The change in resistance of some materials due to heating was initially discovered by Sir Humphrey in 1821 and later on Sir William Siemens experimented on platinum and observed that resistance of platinum increases as its temperature increased. Finally, in 1932, C. H. Meyers developed first platinum based RTD [8]. Since then, RTDs are being used to measure temperature by correlating the resistance of the RTD element with temperature. RTD is basically a resistive element that exhibits a resistance-temperature relationship given by the following equation for a limited range:

$$\frac{\Delta R}{R_o} = \alpha(T - T_o),\qquad(2.1)$$

where

ΔR is the change in resistance due to change in temperature;

R_o is the resistance of the sensor at a reference temperature T_o. The reference temperature is usually specified as $T_o = 0\ °C$;

α is the temperature coefficient of resistivity.

The value of temperature coefficient of resistivity can be controlled by doping i.e. carefully introducing impurities into the metal. The impurities introduced during doping become embedded in the lattice structure of the metal and it results in a different resistance vs. temperature curve.

RTDs are temperature sensors generally made from a pure (or lightly doped) metal whose resistance increases with increasing temperature (positive resistance temperature coefficient). Most RTD elements consist of a length of fine coiled wire wrapped around a ceramic or glass core. The element is usually quite fragile, so it is often placed inside a sheathed probe to protect it. The resistive element of RTD is laid down on a ceramic substrate as a zigzag metallic track a few micrometers thick. Laser trimming of the metal track precisely controls the resistance. Large reduction in size with increased resistance that this construction allows, gives a much lower thermal inertia, resulting in faster response and good sensitivity. They are slowly replacing the use of thermocouples in many industrial applications below 600 °C, due to higher accuracy and repeatability.

Sensing elements are available in a variety of metals, which include copper, platinum, nickel, tungsten and nickel/iron alloy. The most widely used sensor consists of a high-purity (99.99 %) platinum wire wound about a ceramic core and sealed in a ceramic/glass capsule. Platinum is a superior material because of its resistance to contamination and corrosion, and being mechanical and electrical properties are stable over long periods of time. Hence, drift is usually less than 0.1 °C when such a sensor is used at its upper temperature limit. The most popular RTD is the platinum film PT-100, with a nominal resistance of 100 Ω at 0 °C and its sensitivity is 0.39 Ω/°C. Platinum has stability over a wide temperature range (-270 °C to 850 °C) and fairly linear resistance characteristics. Tungsten RTDs are usually used for very high temperature applications. High resistance (1000 Ω) nickel RTDs are also available. If the RTD element is not mechanically stressed and is not contaminated by impurities, these are stable over a long period, reliable and accurate. Platinum RTDs are rarely used above 650 °C in industrial applications. At temperature above 450 °C it becomes increasingly difficult to prevent the platinum from becoming contaminated by impurities from the metal sheath of the sensor.

2.2.3 Thermistors

Thermistor [9] is a temperature sensitive resistor like RTD but its behavior is opposite to RTDs. Its resistance decreases with increase in temperature which is opposite to that of RTDs. Thermistor is a temperature-sensitive resistor made of semiconducting materials, such as oxides of nickel, cobalt, or manganese and sulfides of iron, aluminium, or copper. The type and composition of semiconductor oxides used govern the resistance value and temperature coefficient of thermistor. The commonly used thermistors exhibit negative temperature coefficient and high degree of sensitivity to small changes in temperature, typically 4 %/ °C. Thus, the resistance of thermistors decreases with an increase in temperature as given by the following relationship:

$$\log_e\left(\frac{\Delta R}{R_o}\right) = \beta\left(\frac{1}{\theta} - \frac{1}{\theta_o}\right), \qquad (2.2)$$

where

R   is the resistance of the thermistor at absolute temperature θ in K;

R_o is the   resistance of the thermistor at reference absolute
         temperature θ_o in K;

β is the material constant that ranges from 3000 to 5000 K.

Their accuracy is typically ten times better than thermocouple but not
as accurate as RTDs. Thermistors are non-linear devices, which is one
of their biggest limitations. These may be used over typical temperature
ranges of –80 °C to 250 °C. They cannot be used beyond 300 °C
because it starts behaving as poor conductor. Below -50 °C the
resistance of thermistor approaches that of a poor insulator (several
MΩ), hence, –100 °C is the lower practical limit of thermistor.
Thermistor is most linear in the cold region say below 0 °C and its
sensitivity is also highest in this region. The limitations of non-linearity
can be overcome by modeling the nonlinearities with quadratic
equations extending the operating temperature range. These exhibit a
high resistance, typically 3 kΩ, 5 kΩ, 6 kΩ and 10 kΩ at 25 °C.

Thermistors are available in wide range of shapes in the form of beads,
discs, rods and probes that can be easily manufactured. They exhibit a
fast thermal response due to their small size, but can be quite fragile as
compared to RTDs. Moreover, self-heating problem is more in
thermistor because of their high resistance. In some cases the thermistor
is coated with a protective material like glass to prevent oxidation.

Due to their fast response time, high sensitivity they are most suitable
for a single point measurement for control applications such as switch
on/off of heater or alarming equipment at a particular temperature
which can be measured very accurately and quickly. The self-heating
characteristic is also useful in control application such as level
indicating alarm. Thermistor can be fitted at a suitable height so as to
dip in the liquid. If level of the liquid falls below this height, its self-
heating is sufficient to cause a current flow (due to rapid decrease in
resistance) and alarm can be activated. When it is submerged in liquid,
its self-heat is dissipated in that liquid and keeps the thermistor cool,
keeping resistance high enough to restrict the current below a threshold.

2.2.4 Integrated Circuits (ICs) Temperature Sensors

The integrated circuit (IC) temperature sensor [10-11] is a
semiconductor device that provides an output current that is

proportional to temperature when an input voltage is applied across its terminals. Integrated Circuit temperature sensor is one of the innovations in thermometry and is by far the most inexpensive. They are basically silicon-based sensing circuits with either analog or digital outputs. The important features are temperature range up to 150 °C, low cost, excellent linearity, and additional signal conditioning, comparators, and digital interfaces. The output from IC temperature sensor can be current or voltage, which is linearly proportional to absolute temperature. Typical values of output are 1 μA/K and 10 mV/K, which is the highest output for the amount of temperature change. Digital format of output is also available, which allows it to be interfaced with microprocessor directly. Fig. 2.3 shows a typical electrical diagram of an IC temperature sensor.

AD592 temperature sensor from M/s Analog Devices is a two terminal monolithic integrated circuit that provides an output current that is proportional to absolute temperature. The transducer acts as a high impedance temperature dependent current source of 1 μA/K for a wide range of supply voltage. This IC can be applied in applications between -25 °C to +125 °C. Expensive linearization circuitry, precision voltage references, bridge components, resistance measuring circuits and cold junction compensation are not required in IC based temperature sensors.

LM35 series are precision integrated-circuit temperature sensors, whose output voltage is linearly proportional to the Celsius (centigrade) temperature. LM35 has a distinct advantage over linear temperature sensors calibrated in Kelvin, as the user is not required to subtract a large constant voltage from its output to obtain convenient centigrade scaling. Since it draws only 60 μA from its supply, it has very low self-heating, less than 0.1 °C in still air. LM35 is rated to operate over a temperature range of -55 to +150 °C, while LM35C is rated for a -40 to +110 °C range. Similarly, LM135 is another precision temperature sensor, which operates as a two terminal zener, and the break down voltage is directly proportional to the absolute temperature at 10 mV/K. The temperature measurement range of LM135 is -55 to +150 °C.

Its disadvantages include its temperature range is less than 200 °C, it requires a power supply, it is slow to react to temperature changes and is self-heating. Table 2.2 shows the comparison of IC temperature sensors.

Supply Voltage

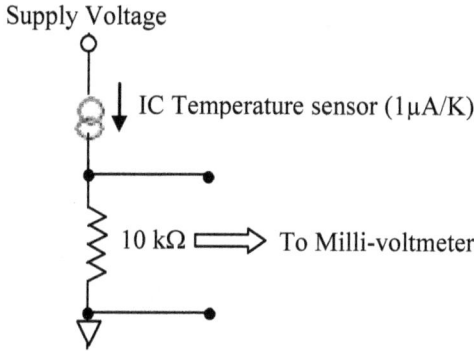

Fig. 2.3. Electrical diagram of an IC temperature sensor.

Table 2.2. Comparison between different temperature sensors.

S. No	Characteristics	Platinum RTD	Thermo-couple	Thermistor	Temperature ICs
1	Active material	Platinum	Two dissimilar metals	Metal oxide ceramic	Silicon
2	Temperature range	-200 °C to 850 °C	-270 °C to 2800 °C	-100 °C to 500° C	-55 °C to 150 °C
3	Accuracy	Best	Moderate	Moderate	Low
3	Linearity	Excellent	Moderate	Poor	Excellent to Moderate
4	Response time	Moderate	Fast	Fastest	Low
5	Stability	Excellent	Poor	Moderate	Moderate
6	Sensitivity	0.39 %C^{-1}	40 μV C^{-1}	-4 % C^{-1}	10 mV C^{-1}
7	Relative sensitivity	Moderate	Highest	Low	Moderate
7	Changing parameter	Resistance	Voltage	Resistance	Voltage
8	Cost	Moderate	Low	Low to moderate	Low
9	Ruggedness	Moderate	High	Low	Moderate

2.2.5 Infrared/Optical Pyrometers

Infrared temperature sensors and optical pyrometers [12] are non contact type temperature measurement devices, which are required in applications with moving parts or where the part may be contaminated if a temperature probe is placed directly on it or object is contained in vacuum or other controlled atmosphere. There are many varieties of infrared temperature sensing devices available today, including configurations designed for flexible and portable handheld use.

We know that all the objects emit energy if their temperature is above absolute zero and the amount of energy increases as the temperature of object increases. The infrared pyrometer receives the infrared light emitted from a source whose temperature is being measured. The most basic design consists of a lens to focus the infrared (IR) energy on to a detector, which converts the energy to an electrical signal that is followed by operational amplifier to amplify the signal. This signal can be directly interfaced either to display device or it can be sent to signal conditioner to convert it in to 4-20 mA or 0-10 V signal for remote application. Fig. 2.4 shows a general schematic of hand held infrared pyrometer. As such, the infrared thermometer is useful for measuring temperature under circumstances where thermocouples or other probe type of sensors cannot be used. Typical temperature range of commercially available optical pyrometer is 700 °C to 350 °C with ± 0.5 % of accuracy.

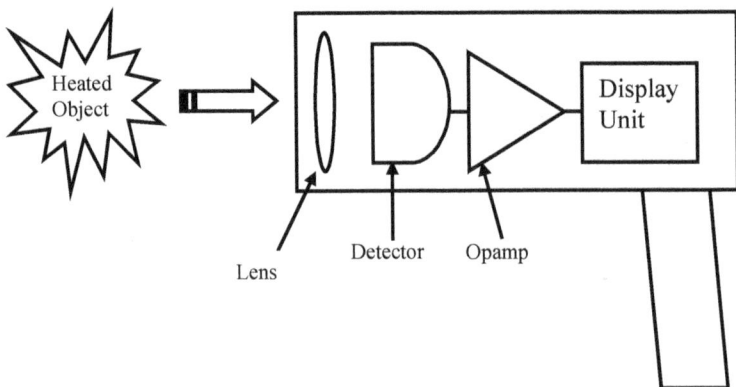

Fig. 2.4. Infrared handheld pyrodetector.

2.3 Selection of Temperature Sensors

Temperature is one of the key parameters, which needs to be measured in chemical laser systems at various locations. Temperature monitoring is also important for measurement, control and optimization of various critical derived parameters of a given system, which may require special techniques that are temperature dependent. These will be discussed in Chapter 3, where we discuss various specialized diagnostics setup.

As discussed in previous section, thermocouples, RTDs, thermistors and integrated temperature circuits are the most common examples of temperature sensors. The electrical parameters of these temperature sensors vary in a particular fashion with temperature. Depending upon the application, one needs to carefully select the sensor type. At the outset it is prudent to evaluate the temperature range over which we want the temperature sensor to function.

In case of CO_2 GDL, measurement temperature range is from 1700-1800 K whereas in HF/DF laser, temperature to be measured is also of the same order. The accuracy and time response requirements are $\pm$ 1 °C and of the order of few milliseconds respectively. The temperature measurement ranges for COIL are from −50 °C to +50 °C for BHP supply system and from room temperature to 150 °C for iodine supply system. The accuracy and time response requirements are 0.2 °C and a fraction of a second respectively.

In case of, HF/DF and CO_2 GDL, integrated temperature circuits and RTDs cannot be used since the measured temperature exceeds their operating ranges. Thermistor are highly nonlinear, hence it is not good from measurement point of view. Therefore, *thermocouple* is best suited because it covers the entire temperature range with fast response time and good linearity. R–type thermocouple may possibly be utilized.

In case of COIL, *RTD Pt-100* (RS component part number 236-4299 or equivalent) may be selected for the measurement of temperature. Since this type of sensor not only meets the requirements of measurement range, accuracy and response time but also it exhibits good sensitivity, stability and linearity. As explained in previous section, platinum resistance temperature detectors offer excellent accuracy over a wide temperature range (−200 to 850 °C). Also, unlike thermocouples, RTD sensors do not require any special cables for the transmission of the information. Other alternatives such as thermistor cannot be used because it is highly nonlinear. Further, use of integrated temperature circuit is also not viable due its lack of compactness thereby leading to difficulty in mounting, and also it barely covers the required temperatures range. Another of its limitation is its incompatibility with a corrosive medium such as BHP, chlorine and iodine. The typical technical specifications of RTD (Pt-100) required for COIL operation are listed in Table 2.3.

Table 2.3. Technical specifications of RTD Pt-100 temperature sensor for COIL application.

Pt-100 sensor probe	Stainless steel (compatible with corrosive medium)
Operating range $(0\,^{0}C)$	-50 to 150
Accuracy at $0\,^{0}C$	± 0.1
Probe length (mm)	250
Cable length (m)	2
Cable insulation	PTFE

2.4 Pressure Sensors

Pressure sensors [13-14] are other most extensively used sensors (after temperature sensor) for industrial applications. The term pressure is defined as the "force applied to per unit area" and can be written mathematically as:

$$P = \frac{F}{A},\qquad(2.3)$$

where F is the magnitude of force in Newton (N) and A is the area in square meters (m^2) over which force is applied. In SI, the unit is 'Nm^{-2}' which is known as Pascal. However, in British units, pressure is expressed as pound per square inch (psi). '*Torr*' and '*bar*' are another units of pressure, which are very commonly used. Atmospheric pressure is used as a reference for most of the units of measurement. The magnitude of pressure produced by the column of atmosphere at the sea level is 14.7 psi or 760 torr.

$$1\,Atm = 14.7\,psi = 760\,torr = 1.01325 \times 10^5\,Pa$$

Pressure measurement can principally be divided into three types:

Absolute pressure: It is measured with respect to vacuum or absolute zero as a reference. For example, pressure measurement by barometer when one of its ports is connected with vacuum.

Differential pressure: It is the difference between two pressures. It is widely used to estimate mass flow rate across the orifice (explained in detail in flow measurement session later in this chapter)

Gauge pressure: It is measured with respect to atmospheric pressure. It is used in measurement of pressure in tiers of automobiles.

Pressure measurement methods have been known for centuries. U-tube manometers were among the first mechanical pressure indicators. Originally, these tubes were made of glass and scales were added later on to provide more accurate measurements. But manometers are large, cumbersome, and not well suited for interfacing with automatic control loops.

Therefore, other methods for pressure measurement have been developed which are very accurate, reliable, stable and compact, which provide electrical output essential for remote monitoring and control applications.

Most pressure transducers are based on the movement of a diaphragm [15] mounted across a pressure differential. Some of the key and widely used pressure transducers are capacitance gauge [16-18], strain gauge [19] and piezoresistor [20] type etc., where the movement is sensed through the electrical signal produced either directly or with the help of bridge circuit (signal conditioning discussed later in Chapter 3). The above three enumerated pressure sensors are discussed in greater details in the following sections.

2.4.1 Capacitance Sensors

Capacitance pressure transducers [16-17] were originally developed for use in low vacuum research. The capacitance change results from the movement of a diaphragm element. The diaphragm is usually metal or metal-coated quartz and is exposed to the process pressure on one side and to the reference pressure on the other. The capacitance pressure sensor consists of a target plate and a second plate known as the sensor head. The two plates are separated by an air gap of thickness d and form the two terminals of a capacitor, which exhibits a capacitance C given by

$$C = \varepsilon \frac{A}{d},$$ (2.4)

where

ε is the permittivity (F/m);

A is the area of sensor head (m^2);

d is the separation distance between plates (m).

It is clear from the above relation that capacitance can be varied by altering any one or all of the three variables. However, in case of capacitance based pressure sensors, separation distance *(d)* between two plates of capacitor is varied due to variation in pressure. If the separation between the two plates is changed by an amount Δd, then the capacitance C becomes

$$C + \Delta C = \varepsilon \frac{A}{d + \Delta d}$$ (2.5)

Taking a ratio of Eq. (2.5) to Eq. (2.4) subsequently reorganizing we may write,

$$\frac{\Delta C}{C} = \frac{-\Delta d / d}{1 + \Delta d / d}$$ (2.6)

It is evident from Eq. (2.6) that $\Delta C/C$ is a nonlinear function of $\Delta d/d$, which is present in the denominator as well. In order to avoid this non linearity, the change in impedance can be measured. We know that

$$Z_c = -\frac{j}{\omega C}$$ (2.7)

With a capacitance change of ΔC,

$$Z_c + \Delta Z_c = -\frac{j}{\omega}\left(\frac{1}{C + \Delta C}\right)$$

$$(2.8)$$

One can also rewrite the above relation as:

$$\frac{\Delta Z_c}{Z_c} = \frac{-\Delta C/C}{1 + \Delta C/C}$$

$$(2.9)$$

Substituting for $\Delta C/C$ from Eq. (2.6) in (2.9) we get,

$$\frac{\Delta Z_c}{Z_c} = \frac{\Delta d}{d}$$

$$(2.10)$$

From above relation, it is clear that the capacitive impedance Z_C is a linear function of d and that methods of measuring ΔZ_C will permit extremely simple plates to be utilized as a sensor to measure the displacement Δd.

The capacitance sensor has several advantages; it is non-contact type allowing it to be used with any target material, it is extremely rugged capable of withstanding high shock loads (5000 g) and intense vibratory environments, can work over wide pressure range varying from high vacuum of the order of few micron to several tens of MPa.

Capacitance sensors are often used as secondary standards, especially in low-differential and low-absolute pressure applications. Stainless steel is the most common diaphragm material used. However, for corrosive service, high-nickel steel alloys such as Inconel or Hastelloy, give better performance. Tantalum also is used for highly corrosive, high temperature applications.

Capacitance sensors are suitable for low-pressure ranges typically from 10^{-2} to 1000 torr with 0.01 % to 0.1 % accuracy and they are highly stable. But they are not suitable for the application of corrosive gases and humid environment. Also, these sensors require warm up time of about one hour for achieving high accuracies and are not ideally suited for the application in corrosive and humid environment.

2.4.2 Strain Gauge Sensors

A strain gauge [19] (also strain gage) is a device which consists of very fine metallic wire that is attached to a surface by a suitable adhesive, such as cyanoacrylate. When it is subjected to pressure, its electrical resistance is changed which is given by the relation below,

$$R = \rho \frac{L}{A},$$

(2.11)

where ρ resistivity of material, L is the length of the wire and A is cross section of the wire. This resistance change, usually measured using a Wheatstone bridge, is related to the strain by the quantity known as the *gauge factor (GF)*. *GF* is basically the ratio of the relative amount of change in resistance to the relative change in the length (i.e. strain) which is given by the relation,

$$GF = \frac{\Delta R / R_G}{\varepsilon}$$

(2.12)

$$\varepsilon = \frac{\Delta L}{L}$$

(2.13)

where

ΔR is the change in resistance caused by strain;

R_G is the resistance of the non deformed gauge;

ε is the strain;

ΔL is the change in length on application of force;

L is the original length without application of force.

Gauge factor is a dimensionless value, and higher the value of GF, the more sensitive the strain gauge. Strain gauges are available commercially with nominal resistance from 30 ohm to 3000 ohm and typical value of gauge factor ranges between 2 to 4.5. The strain gauges are available for the pressure ranges as low as few torr to as high as 150 bar with accuracies in the range of 0.1 % to 0.25 %

2.4.3 Tenso-resistive Type (Piezoresistive Strain Gauge) Sensors

Semiconductor materials like silicon exhibit change in resistance in response to force or pressure applied. This phenomenon is called Piezoresistive effect. Silicon is mostly preferred due to its good elasticity and return to its original shape on release of applied pressure. A Piezoresistive Pressure Sensor [20] contains several thin wafers of silicon embedded between protective surfaces. The surface is usually connected to a Wheatstone bridge, a device for detecting small differences in resistance corresponding to change in pressure.

Semiconductor strain gauges are superior to metal semiconductors in a number of ways. They have a very high gauge factor (typically ~30 as compared to ~2 in case of metal strain gauge). Hence they are useful in the measurement of very small strain of the order of 0.01 μm. Moreover, semiconductor strain gauge exhibits very low hysteresis i.e. less than 0.05 %.

A typical piezoresistive pressure sensor consists of a diaphragm onto which four pairs of silicon resistors are bonded. These sensors are diaphragm type, in which a monocrystalline sapphire plate with silicon film tensoresistors (silicon-on-sapphire-structure) is connected fast to metal diaphragm of tensotransducer. Tensoresistors [21] are connected into a bridge circuit. Deformation of measuring diaphragm (deformation of tensotransducer's diaphragm) causes proportional resistance change of tensoresistors and misbalance of bridge circuit (will be discussed in Chapter 4). Electrical signal from bridge circuit output can be fed to microprocessor based electronic module, where it may be transformed into unified current output (4-20 mA).

The tensoresistor type diaphragm pressure sensors are very accurate (typically 0.25 %) and capable of operating in wide range of temperatures (-40 to 70 °C) and covers wide ranges of pressure measurement (few torr to few hundred bar) as well. They are highly stable for longer period (2-3 years) and support corrosive medium because of special kind of sensor sealing (PTFE/Teflon).

In addition, there are pressure sensors such as thermal conductivity based pressure sensors (*Pirani gauge*: 1 torr to 10^{-4} torr), cold cathode ionization gauges (*Penning gages*: 10^{-4} torr to 10^{-7} torr) for low vacuum measurement.

Pirani gauge works on the principle of thermal conductivity of gas, which is proportional to its pressure (at low pressures). The change in resistance of a filament wire occurs due to change in temperature, which changes with variation in the pressure being measured. The thermal conductivity of each gas is different, so the gauge has to be calibrated for the individual gas being measured. Pirani gauge is linear in 10^{-2} torr to 10^{-4} torr range. Pirani gauges are inexpensive, convenient, and reasonably accurate. The limited reproducibility is main drawback of pirani gauge.

In penning gauge, gas discharge is produced in cross electric and magnetic fields. It measures the flow of charged gas particles (ions), which varies due to density changes to measure pressure. Robust and highly sensitive penning gauges have drawback that they are prone to contamination especially in case of using them in systems fitted with oil-vapor diffusion pumps. From the viewpoint of chemical gas lasers, both pirani and penning gauges do not fall in the vacuum pressure ranges (few torr to few tens of torr) of gas lasers. Hence they are normally not used in chemical gas laser applications. Table 2.4 shows the comparison of capacitance, strain gauge and piezoresistive type pressure sensors.

Table 2.4. Comparison between different pressure sensors.

S. No	Characteristics	Capacitance pressure sensor	Strain gauge pressure sensor	Piezoresistive strain gauge pressure sensor
1	Active material	Thin plates of metal or metal coated quartz	Metal or metal-coated quartz	Semiconductor (Silicon)
2	Pressure range	10^{-2} to few tens of MPa	Few torr to 150 bar	Few torr to few hundred bars
3	Accuracy	0.01 % to 0.1 %	0.1 %	0.25 %
4	Response time	Slower	Fast	Faster
5	Warming time	Required	Not required	Not required
6	Corrosive medium compatibility	Poor	Good	Best
6	Sensitivity	Highest	High	Higher
7	Changing parameter	Capacitance	Resistance	Resistance
8	Electronics for Signal conditioning	More complex	Simple	Simple
9	Cost	High	Moderate	Low to moderate

2.5 Selection of Pressure Sensors

Pressure is another physical variable, which provides great deal of insight whether the supply systems are functioning as per design and also regarding the various phenomenon occurring inside the gas dynamic tunnel of a chemical gas laser. The selection of the pressure sensors needs to be corresponding to the operating range, desired response time, accuracy and compatibility with the flowing medium.

In COIL, the typical pressure ranges are 0.1 to 100 torr for various locations in the gas dynamic tunnel, 1 to 1000 torr in the supply lines for liquid reagent (i.e. BHP) and gases such as chlorine and iodine and 1 to 10 bar for the nitrogen gas supply lines. The sensors are required to work in corrosive environment (iodine, chlorine and BHP) with the ambient temperature ranging from ~ -20 °C to ~70 °C.

Capacitance gauges (model no. CMR 262 from M/s Pfeiffer) may be used since they have high accuracy (0.2 %). However, these sensors require warm up time of about one hour for achieving these accuracies. Moreover, the performance is adversely affected by the presence of water vapor and hazardous gases such as Cl_2 and I_2, which are usually encountered in COIL application.

Also, in case of iodine line pressure measurement, the ambient temperature may be of the order of 70 °C. Hence, tensoresistor type "Metran-22 series" pressure transmitters from M/s Metran [21] are amongst the best suitable for corrosive environment and satisfy COIL application necessities.

The SMC make pressure sensor (Model number ISE 30) are strain gauge sensors capable of measuring pressure in the range of 0 to 10 bar. This sensor provides 4-20 mA signal for acquisition through the analog input channel of the data acquisition card.

Typical fuel supply pressure range for CO_2 GDL and HF/DF laser are 0 to 100 bar and 0 to 10 bar respectively, whereas laser chamber pressure is in the range of 0.1 to 100 torr. The sensors are required to work in corrosive environment since both toluene (used in GDL) and fluorine (in HF/DF) are both corrosive. Also, for combustion driven systems such as GDL and HF/ DF the operating temperature for the pressure sensors are relatively higher.

Higher pressures of 0 to 100 bar as encountered in supply lines and combustor of CO_2 GDL may be measured using strain gauges/tensoresistor type pressure gauges. There are several suppliers manufacturing these sensors such as American sensor technologies Inc (Model number AST 4000 series), USA, Metran Inc.(Model number 2000 series), Russia, Honeywell Inc., USA to name a few. The typical technical specifications for various chemical gas lasers are shown for reference in Table 2.5. It is desirable that sensors should operate at 24 V_{dc} excitation and provide 4 – 20 mA output from data acquisition point of view (explained in Chapter 5)

Table 2.5. Typical technical specifications of pressure sensors required by Chemical gas lasers.

Parameters	Ranges
Absolute pressure	1-100 torr 1-1000 torr 1-10 bar 1-200 bar
Accuracy %	0.25
Operating voltage (V_{dc})	24
Output signal (mA)	4-20
Ambient temperature ($^\circ C$)	-40 to 70

2.6 Level Sensors

Chemical gas lasers employ liquid fuel for their operation. For example liquid toluene in CO_2 GDL and liquid basic hydrogen peroxide in COIL is stored in tank. It is necessary to measure the liquid level of the tank. There are several methods for measurement of liquid level in the tank e.g. hydrostatic pressure (differential) based level sensor, RF capacitance and Ultra sonic liquid level sensor etc. The simplest method is to connect a graduated transparent tube to the bottom of the tank. The liquid level height in the tank is same as the liquid height in the pipe and one can physically see the level through this pipe. However for remote measurement, this method cannot be used. This method does not provide any electrical output for transmitting the signal to remote location for signal monitoring by data acquisition system. This crude method is not suitable for chemical gas laser applications.

2.6.1 Hydrostatic (Differential) Pressure Based Liquid Level Sensor

There are several methods that may be used but differential pressure measurement is apparently the most popular and easiest technique, which may be used. It basically makes use of a differential pressure detector installed at the bottom of the tank whose level is to be detected. The liquid inside the tank creates pressure which is comparatively higher than the reference atmospheric pressure. This pressure comparison is performed via the Differential pressure detector. Fig. 2.5 shows the concept for liquid level measurement inside a tank.

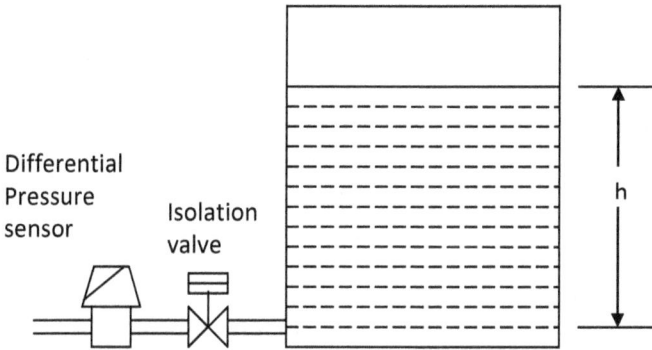

Fig. 2.5. Concept of liquid level measurement.

The level of liquid *h* in the tank is related with the differential pressure *P* (hydrostatic pressure) using the relation:

$$P = \rho g h, \qquad (2.14)$$

where ρ is the density of the liquid and g is the gravitational acceleration (9.8 ms^{-2}). Once the height is known, one can estimate amount of liquid inside the tank if dimensions of tank are known.

In case of tanks which are open to the atmosphere, only high pressure ends of the differential pressure transmitter are needed to be connected whereas the low pressure end of the differential pressure transmitter is expelled into the atmosphere. Hence, the differential pressure happens to be the hydrostatic head or weight of the fluid contained in the tank. The highest level detected by the differential pressure transmitter usually depends upon the maximum height of fluid above the transmitter, whereas the lowest level detected is based upon the position where the transmitter is attached to the tank or vessel. Differential pressure based level sensors [22] have advantage of easy mounting on to the surface of the vessel or tank, further maintenance and testing may be carried out by isolating them using block valves.

2.6.2 RF Capacitance Liquid Level Sensors

RF (radio frequency) technology uses the electrical characteristics of a capacitor, in several different configurations, for level measurement. It is commonly referred to as RF capacitance [22] or simply RF method and suited for detecting the level of liquids in a vessel.

An electrical capacitance (the ability to store an electrical charge) exists between two conductors separated by a distance, d, as given by relation (2.4). The first conductor can be the vessel wall (plate 1), and the second can be a measurement probe or electrode (plate 2). The two conductors have an effective area, A, normal to each other. Liquid whose level is to be measured can be in between the conductors. The amount of capacitance here is determined not only by the spacing and area of the conductors, but also by the electrical characteristic dielectric constant (or permittivity) of the liquid to be measured.

In order to apply this formula to a level-measuring system, one must assume that the process material is insulating, which, of course, is not always true. A bare, conductive, sensing electrode (probe) may be inserted down into a tank to act as one conductor of the capacitor. The metal wall of the tank acts as the other. If the tank is nonmetallic, a conductive ground reference must be inserted into the tank to act as the other capacitor conductor.

When there is no liquid inside the tank, the insulating medium between the two conductors is air and it will have some capacitance value. As the level rises in the tank, a change in capacitance will occur between the sensing probe and ground. This capacitance is measured to provide a direct, linear measurement of tank level.

From electrical measurement point of view, another approach may be to measure impedance of the medium which offers improved reliability and wider range of uses. In RF or AC circuits, impedance, Z, is defined as the total opposition to current flow:

$$Z = R + \frac{1}{j(2\pi fC)}, \qquad (2.15)$$

where

R is the resistance in Ohms;

j is the $\sqrt{(-1)}$;

f is the measurement frequency (radio frequency for RF measurement);

C is the capacitance value.

An RF impedance level-sensing instrument measures this total impedance rather than just the capacitance. In some cases, the process material tends to build up a coating on the level-sensing probe. In such cases, which are not uncommon in level applications, a significant measurement error can occur because the instrument measures extra capacitance and resistance from the coating buildup. As a result, the sensor reports a higher, and incorrect, level instead of the actual tank level.

The sensor contains no moving parts, is rugged, simple to use, and easy to clean, and can be designed for high temperature and pressure applications. A danger exists from build-up and discharge of a high-voltage static charge that results from the rubbing and movement of low dielectric materials, but this danger can be eliminated with proper design and grounding.

2.6.3 Ultra Sonic and Sonic Liquid Level Sensors

Both ultrasonic and sonic level sensors [22] operate on the basic principle of using sound waves to determine fluid level. The sonic level sensors takes advantage of the principle that the speed of the sound waves traveling through air or gas can be measured and timed. In this method, a top-of-tank mounted transducer transmits waves downward in bursts onto the surface of the material whose level is to be measured. The sensor has a transmitter and receiver mounted on the same head. An electrical pulse is applied to a piezoelectric crystal that causes it to vibrate against a diaphragm and produces sound energy that travels in the form of a wave at the established frequency and at a constant speed (v) in a given medium. The sound waves are directed at the level of the liquid being measured and they will be reflected back to the sensor. The medium is normally air over the material's surface but it could be nitrogen or some other vapor. The sensor measures the time (T) for the bursts to travel down to the reflecting surface and return. This time to the surface and back can be used to determine the level of fluid in the tank using the basic relation:

$$D = Txv \qquad (2.16)$$

The frequency range for ultrasonic methods is ~20–200 kHz, and sonic types use a frequency of ≤ 10 kHz. These level sensors are used for non-contact level sensing of highly viscous liquids. They are also widely used in water treatment applications for pump control and open channel flow measurement.

Ultrasonic level sensors are affected by the changing speed of sound due to moisture, temperature, and pressures. Correction factors must be applied to the level measurement to improve the accuracy of measurement. In this method, proper mounting of the transducer is required to ensure best response to reflected sound. Since the ultrasonic transducer is used both for transmitting and receiving the acoustic energy, it is subject to a period of mechanical vibration known as "ringing". This vibration must attenuate (stop) before the echoed signal can be processed.

It is difficult to compare the capabilities of sensors based on different principles however; a typical comparison which may potentially assist in making a suitable sensor selection for specific application is put forth in Table 2.6.

Table 2.6. Comparison between different liquid level sensors.

S. No	Characteristics	Hydrostatic pressure level sensor	RF Capacitance level sensor	Ultrasonic/sonic level sensor
1	Principle	Hydrostatic pressure	Capacitance/ Impedance	Time of flight
2	Typical range	0 to 60 bar	Few cm to few m	Few cm to few tens of m
3	Accuracy	0.25 %	0.25 %	0.20 %
4	Operating tempe- rature range (°C)	-25 to 100	-25 to 70	-25 to 80
5	Sensor type	Contact type	Contact type	Non contact type
6	Electrical output	4-20 mA	4-20 mA	4-20 mA

2.7 Selection of Level Sensors

Level sensors are required for constant monitoring the level of liquid reagents in case of both GDL and COIL. HF/ DF lasers in general do not employ any critical liquid reagent. As discussed in previous section, there are various methods available for level sensing viz., RF capacitance method, ultrasonic method and differential pressure (hydrostatic) pressure measurement.

In case of chemical gas lasers, one essentially requires an easily realizable, passive and a rugged approach to sensing liquid level. Hence, the differential pressure measurement is the most suitable since other approach such as RF capacitance may lead to interference with an already extensively wired system. The ultrasonic method, on the other hand, requires a clear line of sight from the source to the tank, which may not be readily available.

Considering, an example of measurement of level of BHP in case of COIL, the level sensor should be, resistant to corrosion from interaction with BHP solution, have an operating temperature range of -30 $^{\circ}$C to 40 $^{\circ}$C. The former is required since BHP solution is sometimes used at high negative temperatures to limit the production of water vapor in the system. The high positive limits occur in case of cold run experiments, wherein water is used instead of BHP solution for evaluating the various gas flow conditions in the gas dynamic tunnel.

Metran 43 series level sensors or equivalent are capable of satisfying the various level range requirements for typical COIL operating temperature and are also corrosion resistant. Moreover, the output of these sensors is 4-20 mA and from implementation point of view, this sensor is very much similar to Metran pressure sensors. So, from user's point of view, all the external electrical connections and design formulas are same as in case of Metran 22 series pressure sensor. Table 2.7 depicts typical technical specification of level sensor required in COIL.

In case of COIL, the BHP tank operating pressure is atmospheric and the decrease in the level of BHP with run time gives a clear indication of the BHP volume flow rate as well for the known geometry of the tank. Hence, a proper calibration of the sensor for the liquid (both BHP and water) and different heights of the tank enables the determination of the liquid flow rates.

Table 2.7. Typical technical specifications of level sensor for COIL.

Parameters	Typical value for COIL
Hydrostatic level pressure (kPa)	10 kPa
Accuracy %	0.25
Operating voltage (Vdc)	24
Output signal (mA)	4-20
Ambient temperature (°C)	-40 to 70

In COIL experiments, amount of liquid level inside the tank can be utilized to estimate the flow rate of liquid BHP transferred from the tank to the singlet oxygen generator during the run. This enables in ascertaining whether the SOG is functioning for the designed BHP flow rates or not. This is central to trouble shooting as well since a lower flow rates clearly indicates a blockage in the flow due to particle deposition on injector plates blocking the injection holes and a higher flow rates is indicative of wear of the injector plates. Hence, this parameter is essential for optimization of process of singlet oxygen generation, which is directly related to flow of BHP. The level of the solution in BHP tank is acquired using the level sensor and utilized for the estimation of flow rates passing through the generator by measuring the rate of decrease in level of solution in the tank.

A typical liquid level reduction in the preparation tank in a cold run experiment (water instead of BHP) of small scale COIL is shown in Fig. 2.6. This reduction in liquid level or slope of the curve gives an idea of the liquid flow through the singlet oxygen generator, an important parameter to study the SOG/ laser performance. In the present system, the water flow rate during cold run is observed to be 0.5 $l\ s^{-1}$ and is crosschecked with the collected water at the receiving tank in each experiment. The system is calibrated during cold run through several experiments and in order to have the actual value of BHP flow rate in laser power runs, a multiplication factor of ~0.769 is incorporated in to the calculations, which corresponds to the density ratio between water and BHP.

In case of GDL, the measurement of toluene levels is critical since its supply at design flow rate is essential for lasing action. Hence, the required level of toluene should be available in the tank prior to the run.

Hence, measurements using the level sensors may be carried out before tank pressurization and the level sensor may be isolated from tank thereafter using solenoid valve for remote operation. This is essential since tanks during the run are pressurized to a pressure of 60 bar and the operating characteristics of the sensor are pressure dependent and may change due to repeated pressurization.

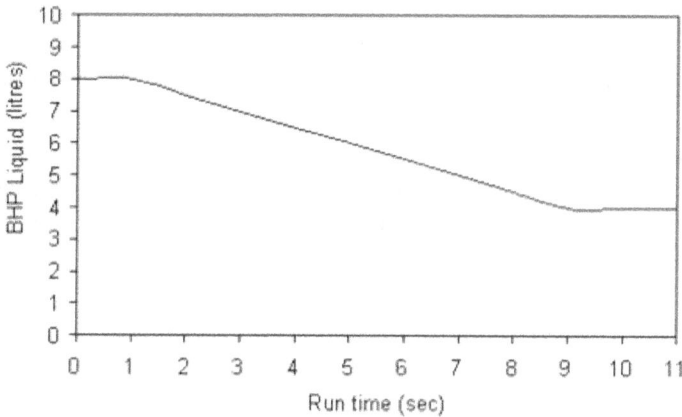

Fig. 2.6. Temporal variation of BHP volume in preparation tank.

2.8 Flow Sensors

Pressure based flow control [23-24] and sensing has a long history in the literature of fluid mechanics. Compact, inexpensive and accurate flow controllers with online measurement and control of flow rates have received a great deal of attention because of their widespread applications in semi conductor process equipment industry, chemical vapor deposition, plasma devices and etch, high and low power gas lasers etc. These applications require wide range of output pressures starting from few Torr to several atmospheres along with a wide range of volumetric flow rates. In addition, online monitoring and regulation of the flow rates are essential in some areas like in high power infrared gas lasers [25-26].

There are several methods for flow measurement [27-31] depending upon the application. There are four main methods for flow measurement enunciated below.

2.8.1 Positive Displacement Flow Meters

In positive displacement flow meters, flow is divided into segments and each segment is measured as the flow occurs. For example in oscillating piston pump, all of the fluid is passed through the known volume of the piston and if number of strokes of piston is known, one can estimate the total volume. This will give us the volumetric flow rates of the fluid by dividing the volume with the time taken in the given number of strokes.

Commonly used positive displacement flow meters are nutating disc, rotating valve, oscillating piston, oval gear, rotating lobes flow meters etc. This method is very accurate and suitable for viscous liquid flow. However, they exhibit high-pressure drop due to its total obstruction in the flow path. They require higher maintenance cost and are not suitable for low flow rates.

2.8.2 Velocity Flow Meters

Velocity flow meters operate linearly with respect to the volume flow rate. Velocity meters have minimum sensitivity to viscosity changes when used at Reynolds numbers above 10,000. Most velocity-type meter housings are equipped with flanges or fittings to permit them to be connected directly into pipelines.

Examples of velocity flow meters are the paddlewheel flow meter, turbine flow meter etc. In paddle wheel flow meter a paddlewheel is inserted in the flow. When flow passes it rotates freely and its speed of rotation is proportional to fluid speed. A magnet is mounted on the each paddle of wheel and sensor detects the motion of the wheel. Turbine flow meter also works in the similar way but with more accuracy. Turbine wheel is mounted directly in the flow. If a fluid moves through a pipe and acts on the vanes of a turbine, the turbine will start to spin and rotate. The rate of spin is measured to calculate the flow. Each vane of turbine wheel has a magnet on it, which is used to generate electric pulse when vane spins under the magnetic pickup. Number of pulses received by detector is used to determine the flow rate of the fluid.

2.8.3 Mass Flow Meters

Mass flow meters are one of the most accurate types of flow meters that can measure the mass flow rates directly. They are suitable for flow measurement of gases as well as other fluids. Coriolis mass flow meter and thermal mass flow meters are two types of flow meters under this category.

Coriolis mass flow meter works on the principle of Coriolis effect to measure the mass flowing though the U shaped tube. In this method, tube vibrates up and down naturally with the help of strong magnet. When fluid starts flowing through the tube, it opposes the up and down movement which results in twist of tube. The amount of twist is directly proportional to the amount of flow through the tube. Coriolis mass flow meter is not sensitive to changes in pressure, temperature, viscosity and density. However, they have drawbacks in terms of larger size and cost as compared to other mass flow meters. There may be a possibility of clogging in pipe, which will be difficult to clean.

Thermal mass flow meter is used to measure mass flow of gases. This method is independent of density, pressure, and viscosity. Thermal flow meters measure the heat carried away from the sensor by the passing flow to determine the mass flow rate. They use a heated sensing element isolated from the fluid flow path where the flow stream causes heat loss from the sensing element. The amount of heat loss is directly proportional to the mass flow rate. The temperature difference is used to calculate the mass flow rates. This method is suitable for measurement of mass flow of clean gases.

2.8.4 Differential Flow Meters

Differential pressure flow meters are based on pressure drop over an obstruction inserted in a flow. Differential pressure flow meters (in most cases) employ the Bernoulli equation that describes the relationship between pressure and velocity of a flow. These devices guide the flow into a section with different cross section areas (different pipe diameters) that causes variations in flow velocity and pressure. By measuring the changes in pressure, the flow velocity can then be calculated. This method may utilize orifice, venturi tube and nozzles device. It would be appropriate to mention here that typically, differential flow meters and more specifically orifice flow meters have been employed and operated for largely incompressible flow regimes.

Table 2.8 shows a comparison between different flow sensors applicable in case of chemical gas lasers.

Table 2.8. Comparison of parameters for various flow sensors.

S. No	Type	Liquid	Gas	Output	Pressure loss	Accuracy % full scale	Cost
1	Orifice	Limited	Good	Square root characteristic	Medium	± 0.5 to ± 2 %	Low
2	Venturi	Limited	Good	Square root characteristic	Low	± 0.5 to ± 3 %	Medium
3	Positive displace-ment	Good	Good	Linear	High	± 0.5 to ± 1 %	Medium
4	Velocity Flow meter	Limited	Poor	Linear	Low	± 2 to ± 5 %	Medium
5	Mass flow meter	Good	Poor	Linear	Low	0.01 % to 0.15 %	High

2.9 Selection of Flow Sensors

In chemical gas lasers, flow rate measurement and control is essential because they are strongly dependent on the mass flow rates of different gases and chemicals.

Nitrogen is a gas constituent common to all chemical gas lasers. For example, COIL operation needs buffer additions for: (1) singlet oxygen transportation termed as primary line, (2) iodine transportation termed as secondary line (which may be again split into two different lines for the iodine flow control termed as secondary main and secondary bypass), (3) mirror blowing and (4) curtain inside the nozzle and also for diffuser startup. Hence, COIL application demands flow rate measurement and control for wide range (few mmole.s^{-1} to few hundred mmole.s^{-1}) with online variation of flow rates.

Similarly, in CO_2 GDL nitrogen is required for mirror blowing, nozzle curtain and diffuser startup. It is also used for pressurization of toluene tank since toluene being combustible is required to be pressurized by a non-oxidizing relatively inert gas.

In case of HF/DF nitrogen is central to laser operation since it is the energy of the nitrogen arc plasma, which dissociates SF_6 in the plenum to generate F atoms, which interact with hydrogen atoms injected close to the nozzle for lasing action. Apart from this nitrogen is also used for mirror blowing, nozzle curtain and for diffuser startup.

Apart from nitrogen, COIL makes use of chlorine gas required for generating the singlet oxygen pumping medium, GDL uses a sufficiently large flow of air which is the oxidizer for combustion and is a rich source of nitrogen needed for efficient energy transfer for lasing purposes, HF/DF lasers use SF_6 that is the source of fluorine required for lasing action.

Hence, it is essential that the gas feed system comprising of the primary gas constituents central for lasers operation and buffer gas such as nitrogen is so designed that we can not only correctly meter the flow rate but can precisely control the flow rates as per need. Also, the gas feed system should control the flow rates and supply pressures of the buffer gas precisely and supply them to various subsystems according to the required sequence. Since the operating time for all the chemical lasers is short and even requiring changes in flow rate during the run itself a fast response for the flow sensors is essential.

There are various ways of measuring the flow rates such as positive displacement flowmeter, velocity flowmeter, mass flowmeter and differential flowmeter already discussed in previous section. The differential flowmeter based on orifices customized for compressible flows appear to be the most suitable candidate for chemical gas lasers applications. Since all other metering devices are capable of measuring the flow rates accurately but it is difficult to couple them with mass flow controllers for flow control and regulation with a desired fast response. Even in differential flow meters there are options of using venturi and nozzle flowmeter, which may be, conditioned for compressible operation regimes. However, for cases where pressure drop is not a limitation orifice based methods is preferred due to ease of fabrication and compactness associated with them.

Therefore, flow rate regulation [27-30] may be achieved using orifices under choked conditions (sonic flow) in which the outlet flow and pressure are proportional to the orifice size and the upstream pressure. Choked flow conditions [31] are achieved by maintaining the downstream (P) and upstream (P_0) pressure conditions, such that the pressure ratio must satisfy the condition:

$$\frac{P}{P_o} \leq \left(\frac{2}{\gamma+1}\right)^{\frac{\gamma}{\gamma-1}} \qquad (2.17)$$

These conditions are occurring normally in chemical gas lasers. Under these conditions, the outlet flow rate is given by [32],

$$m = \sqrt{\frac{\gamma}{R}} \frac{P_o A}{\sqrt{T_o}} \left[\frac{2}{\gamma+1}\right]^{\frac{\gamma+1}{2(\gamma-1)}} C_d \qquad (2.18)$$

where

γ is the specific heat ratio of the gas (=1.4 for nitrogen and 1.67 for helium);

R is the characteristic gas constant;

P_o is the upstream stagnation Pressure;

T_o is the stagnation temperature;

A is the orifice area;

C_d is the discharge coefficient.

In case of gas lasers, fast pressure regulation and time required switching on/off of the system determines the control response time for the system. Further, the accuracy and resolution in pressure regulation determine the same for the flow rates. The line pressure and upstream orifice pressures are monitored using suitable pressure gauges for determining the flow rates.

A typical scheme of operation of a customized choked flow orifices based mass flow controllers [33] is shown in Fig. 2.7 along with the its possible interfacing with data acquisition system.

As stated earlier, the flow rate of gas feed line is directly proportional to the upstream pressure of orifice under choked flow conditions. Hence, the flow rate can be adjusted by controlling the upstream pressure of the orifice. The line pressure can be controlled using an electrical pressure reducer whose output line pressure is a linear function of applied analog input voltage. Thus, in order to control flow rate of gas species the analog output voltage needs to be controlled. An electrical pressure reducer (M/s SMC Pneumatics Make,

ETV 3050 Series or M/s Tescom Make ER 3000 Series or equivalent) may be used for this purpose which has a maximum inlet pressure of 10 bar and outlet regulation from 1 to 9 bar with a response time of about 20 msec. The electrical pressure reducer adjusts the outlet pressure according to the applied analog input voltage (0-10 V) or current (4-20 mA) to the pressure reducer. In case of GDL, higher pressures (like 60 bar in toluene tank) may be achieved by using mechanical ratio loaders after electrical pressure reducer.

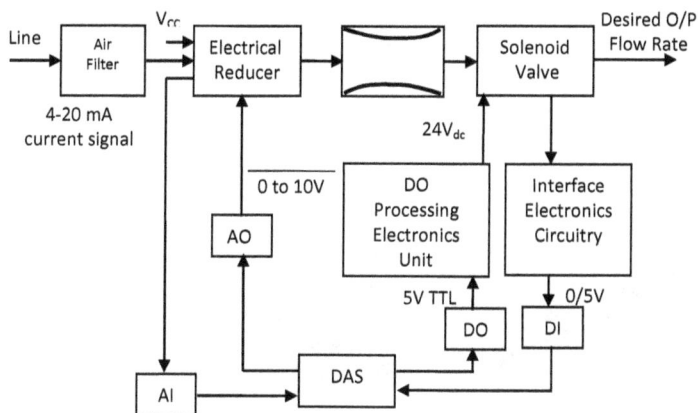

Fig. 2.7. Orifice based gas flow measurement scheme with DAS interfacing.

The required supply voltage to the electrical reducer for its operation is 24 V dc. The 24 V dc is provided by a regulated DC power supply. The analog output channel of the DAS card generates the analog voltage required for the pressure regulation. The electrical pressure reducer also provides 4-20 mA current output as the function of the outlet pressure. This output current that is converted in the range of 0 to 10 V using load resistance as per the requirement of analog input channels of data acquisition system (DAS). This signal is acquired by DAS card and stored for temporal variation for further analysis of chemical gas lasers. This electrical signal is calibrated in terms of pressure and used for the online visualization of the pressure on the graphical user interface (explained in detail in Chapter 5).

The gas may be fed into the system by using an electrical solenoid valve (M/s SMC Pneumatics Make, VXZ 2000 Series or equivalent). The solenoid valve operates at 24 V dc. The solenoid valve controls the supply of gas, whose sequential operation is controlled by the digital output channel of DAS card. The digital input channel of the data acquisition card acquires the on/off status of the solenoid valve. The filter is generally used for the protection of the electrical reducer from foreign particles. The photograph of the typical developed hardware for buffer gas feed of small scale COIL system is shown in Fig. 2.8.

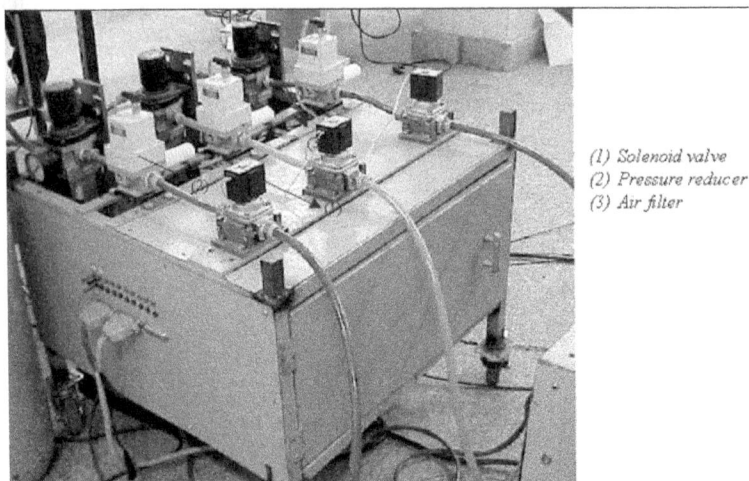

(1) Solenoid valve
(2) Pressure reducer
(3) Air filter

Fig. 2.8. Photograph of gas flow control-DACS interface.

The discharge coefficient (C_d) appearing in Eq. (2.18) of the particular orifice can be determined from the calibration of the system using any standard electronic flow measuring unit for the particular gas species and its typically value is ~0.93-0.95. Fig. 2.9 shows the calibration curve used to calculate coefficient of discharge of an orifice used for nitrogen gas supplied from standard nitrogen cylinders at a temperature of 300 K.

Apart from gas flow sensors, in case of GDL, liquid flow sensor is also required for the measurement of toluene flow rates. The most suitable candidate group for measuring the flow is supposedly the velocity flow meter in which turbine flow meters may be employed or mass flow meters in which Coriolis flow meters could be a good choice. The flow meters grouped under the positive displacement category are associated with significantly high-pressure drops and are therefore not suitable to be employed. Differential flow meter group is also not very suitable on account of size and requiring multiple pressure measurement for correlating with flow rates.

Flow rate Vs Pressure (for Nitrogen)

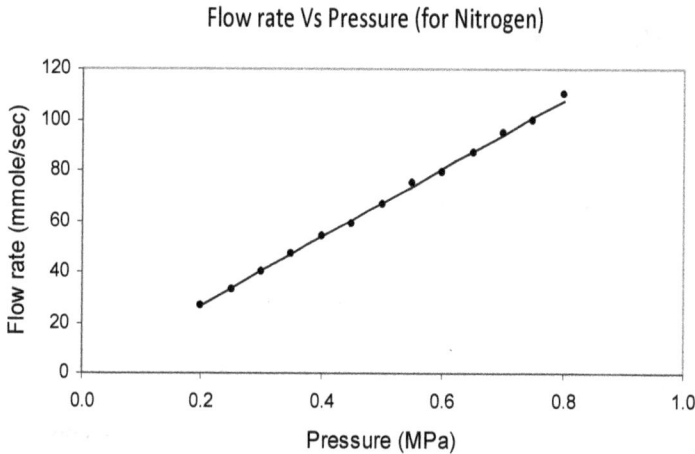

Fig. 2.9. Observed flow rates (filled circles) for different upstream pressure along with theoretical value (line) for a discharge coefficient of 0.93.

Amongst the turbine flow meter and Coriolis flow meter the latter is highly accurate (0.1 %) but is mostly used for low flow rate ranges. It is because for high flow rates its size is too large to be accommodated in chemical gas lasers systems where space in on premium. Thus, for high flow rate conditions turbine flow meters are normally employed since they are relatively compact making a slight compromise in terms of accuracy and stability of output. In CO_2 GDL, Coriolis mass flow meter from M/s Micro Motion, (Model No F100 series with IFT9701 transmitter), USA or equivalent from Bronkhorst (UK) Ltd. can be used to accurately measure the mass flow rates of toluene, which is crucial in combustor operation. The technical specifications of mass flow meter required in CO_2 GDL are shown in Table 2.9.

Table 2.9. Typical technical specifications of mass flow meter for CO_2 GDL.

Parameters	Range
Max mass flow rate (kg/hr)	10000
Accuracy %	± 0.1 % to ± 0.2 % of flow rate
Pressure (bar)	100
Output signal (mA)	4-20
Temperature (^{0} C)	-40 to 100

2.10   Optical Sensors

Optical sensors also commonly termed as photo detectors measure the intensity of the photo signal. Since wavelengths to be measured may vary over a wide range from X – ray to far IR, hence, suitable photo detectors are required which exhibit sufficient sensitivity in the desired range.

There are two basic types of photo detectors, thermal and photon. Thermal detectors utilize thermocouples and thermopiles, which employ heat transfer and temperature, change to sense radiation. These detectors [34] with time constant of the order of 100 ms respond very slowly when compared with photon detectors. Thermal detectors are used effectively to measure light intensity in continuous-wave laser beam, where the intensity of the radiation is extremely high and the intensity of output is constant with time. The photon detectors in general respond directly to the photon flux and exhibit a much quicker response. Both thermal and photon detectors are discussed in detail in the following sections.

2.10.1 Thermal Detectors

The absorption of IR energy heats the detection element in energy or thermal detectors, leading to changes in physical properties which can be detected by external instrumentation Thermal detectors are operated at room temperature and have a wide spectral response. Since the operation of thermal detectors involves a change in temperature, they have an inherently slow response and have relatively low sensitivity compared to photon detectors. The response time and sensitivity of thermal detectors are influenced by the heat capacity of detector structure as well as the optical radiation wavelength. In some applications of thermal detectors, an optical chopper is also needed. The following are examples of energy detectors.

2.10.1.1 Thermocouples / Thermopiles

Thermocouples / Thermopiles are formed by joining two dissimilar metals that create a voltage at their junction. This voltage is proportional to the temperature of the junction. When an IR source is optically focused onto a thermocouple detector its temperature increases or decreases as the incident IR flux increases or decreases.

76

The change in IR flux emitted by the source can be detected by monitoring the voltage generated by the thermocouple. For sensitive detection, the thermocouple must be thermally insulated from its surroundings. For fast response, the thermocouple must be able to quickly release built up heat. This tradeoff between sensitivity of detection and the ability to respond to quickly changing scenes is inherent to all energy detectors. A thermopile is a series of thermocouples connected together to provide increased responsivity. This concept is used for calorimetric based power measurement of high power infrared gas lasers.

2.10.1.2 Pyroelectric Detectors

Pyroelectric Detector is an electromagnetic radiation detector whose operation is based on the pyroelectric effect, that is, the temperature dependence of the spontaneous polarization. These detectors are basically thermal radiation sensors and are based on a special type of materials in which spontaneous polarization changes with temperature. This polarization change gives rise to a voltage across the crystal. The sensing element of a pyroelectric detector is a thin sheet of a pyroelectric, such as triglycine sulfite, barium titanate, or lead titanate, or lithium tantalite with electrodes applied to the surfaces perpendicular to the polar axis. The electrode facing the radiation source has an absorptive coating whose optical properties determine the region of spectral sensitivity of the detector. Since these sensors can only respond to changing thermal scenarios, in order to measure the temperature of a stable radiation source, one requires chopper. These detectors when operated in a chopped system; the fluctuation in the exposure to the source generates a corresponding fluctuation in polarization, which produces an alternating current that can be monitored with an external amplifier. They have the advantage of large dynamic range covering power ranges from 10 nW to 10 W and can be used over a very large wavelength range from fraction of a micron to millimeters wave. However their sensitivity is less as compared to photon detectors.

2.10.1.3 Thermistors / Bolometers

In thermistors, the resistance of the elements varies with temperature. One example of a thermistor is a bolometer. Bolometers function in one of the two ways: monitoring voltage with constant current or monitoring current with constant voltage. Advances in the micromachining of silicon have ushered in the exciting field of microbolometers, capable of measuring temperature changes of 0.1 °C from a power input of 10 nW. A microbolometer consists of an array of bolometers fabricated directly onto a silicon readout circuit. This technology has demonstrated excellent imagery in IR. Although, the performance of microbolometers currently falls short of that of photon detectors, development is underway to close the performance gap. Microbolometers can operate near room temperature and therefore do not need vacuum evacuated, cryogenically cooled dewars. This advantage brings with it the possibility of producing low cost night vision systems for both military and commercial markets.

2.10.2 Photon Detectors

Photon detectors are used more widely as a general-purpose sensor because they are more sensitive and respond more quickly with time. The main photo detectors are vacuum tube photosensors, photomultiplier tube (PMT) and semiconductor photodiodes etc. The vacuum tube photosensors have relatively low sensitivity, but can detect high frequency light variations (as high as 100 MHz to 1 GHz), for an extremely fast response.

The photomultiplier tube (PMT) is the most popular vacuum device. It is similar to a vacuum photodiode with extra electrodes between the photo cathode and anode, called dynodes.

Semiconductor [35] photodiode is a PN junction diode with a transparent window that produces charge carriers at a rate proportional to the intensity of incident light. Thus, photodiodes [34] are usually based on silicon, germanium, indium gallium arsenide and operate by absorption of photons and generate a flow of current in an external circuit, proportional to the incident power. Photodiodes can be used to detect intensities down to 1 pW/cm^2. The semiconductor photodiodes, being small, rugged and inexpensive, have replaced the vacuum-tube detectors in many applications.

Silicon photodiode is a semiconductor with band gap energy of 1.12 eV at room temperature, which corresponds to cut off wavelength of 1100 nm [36]. The band gap energy for Ge is 0.72 eV which corresponds to a maximum cut off wavelength of ~1700 nm [36] although its responsivity drops towards the maxima. Hence, for longer wavelengths (800-1700 nm), InGaAs photo diode can be used [36].

Most commonly used detectors and their typical working ranges are briefly discussed below.

2.10.2.1 Silicon/Germanium Detectors

Silicon and Germanium detectors have inherent advantages of manufacturing due to compatibility with semiconductor production techniques. The use in IR detector stems from variations possible due to doping. In case of silicon, doping with gold gives it an energy band gap of 0.02 eV and cut off wavelength of 5 μm. These figures get altered to 0.05 eV and ~20 μm with phosphorus doping. Similarly the quantum efficiency, electric field and thickness can be varied from 20 % to 40 %, 100 to 500 $V.cm^{-1}$ and 1 to 3 μm respectively. Doping silicon with boron, arsenic, or gallium, for example, introduces different energy levels into the host-material band gap. Electrons can then be knocked off the dopants at energy levels well below the cutoff wavelengths for silicon or germanium, and IR detection at longer wavelengths becomes possible.

2.10.2.2 MCT Detectors

MCT detectors have been the most important semiconductor for mid and long-wavelength (3-30 μm) infrared photo detectors. No single known material surpasses MCT in fundamental performance and flexibility. MCT ($Hg_{1-x}Cd_xTe$) is a combination of mercury telluride and cadmium telluride. Relative concentrations of two molecules i.e. x and 1-x are deliberately adjusted in growth process to obtain desired mixture. This helps to adjust cut off wavelength (maximum wavelength of response). Hence, MCT exhibits extreme flexibility. It can be tailored for optimized detection at any region of the IR spectrum. Limitations of MCT are compositional non-uniformity, difficulty to grow on silicon and fragility. MCT provides better sensitivity, faster response, and lower bias voltage for high-performance applications such as thermal imaging, radiometry, and photoconductivity [37]

2.10.2.3 Indium Antimonide (InSb)

Indium antimonide detectors are most commonly used III-V material providing high performance detectors in the wavelength region from 2 to 5 μm. InSb is a detector material that is very common in single-detector, mechanically scanned units. It is a highly versatile detector, which can be used in cooled and uncooled formats thereby justifying the economics of using it. The material typically offers higher sensitivity as a result of its very high quantum efficiency (80 %-90 %). However, high quantum efficiency is not the most important factor. With InSb, the detectors swamp in a few microseconds, but then the rest of the photons must be dumped. As a result, for most applications there is little benefit to the added quantum efficiency. Another drawback is that InSb infrared FPAs have been found to drift in their non-uniformity characteristics over time and from one cool down to the other, thus requiring periodic corrections in the field. As a result, the system becomes more complex by requiring thermoelectric coolers, and additional electronics. Thus, few manufacturers use InSb FPA detectors for measurement applications. The added complexity of an InSb system is generally warranted in applications where extreme thermal sensitivity is required, for example, long-range military imaging.

2.10.2.4 Ternary Compounds Detectors

Ternary compounds made from III-V elements such as indium gallium arsenide (InGaAs) are already available. They operate in the near-IR region at 1 to 1.8 μm or higher, and no cryogenic cooling is required. Very good performance using thermoelectric cooling can be achieved below 1.8 μm. Linear arrays made from InGaAs elements are also commercially available.

2.10.2.5 Alternate Indium Antimonide Detectors

Alternate Indium antimonide detectors based on InAsSb offer several advantages over InSb detectors. First, the addition of arsenic to the compound material, say ($InAs_{0.80}Sb_{0.20}$), increases the band gap slightly and consequently reduces the maximum detectable wavelength to 5 μm compared to almost 6 μm for InSb. In this case, thermoelectric cooling is sufficient, with advantages such as compactness and reduced cost. When compared to ternary compounds such as MCT, manufacturing is more predictable; in the case of MCT, variation of the band gap is much more sensitive to the relative composition of the elements in the material.

2.10.2.6 Platinum Silicide (PtSi) Detectors

Platinum silicide detectors operate in the short-wavelength region (1-5 μm), have good sensitivity (as low as 0.05°C), and excellent stability. It can be manufactured with semiconductor production techniques, with fairly high detector yields resulting in reasonable costs. Platinum silicide has been desirable for measurement cameras and FPAs because it is a highly stable material that resists drift over time in its responsivity to temperature. One drawback is low quantum efficiency (<1 %). However, modern signal-processing techniques coupled with hybrid construction and CMOS readouts have made PtSi a leading material for use in preventive maintenance and scientific IR imaging environments.

A comparative assessment of thermal and photon detectors, for selection of the most suitable device for photo signal measurement is given in Table 2.10.

Table 2.10. Comparison of parameters for thermal and photon detector.

Parameter	Photon Detector	Thermal Detector
Response time	Fast	Slow
Spectral responsivity	Narrow and selective	Wide and Flat
Sensitivity	High	Low
Operating temperature	Cryogenic	Room
Cost	Expensive	Economical

2.11 Selection of Optical Sensors

In chemical gas lasers, operating wavelengths are 2.7 - 3.4 μm for HF-DF laser, 10.6 μm for CO_2 GDL and 1.315 μm for COIL. Suitable photo-detectors are required for detection of these wavelengths. In addition, it is necessary to develop optical measurement of concentration of the various constituents of the lasing medium, like in case of COIL, lasing medium i.e. iodine, pumping medium i.e. singlet oxygen and also others such as unutilized chlorine, water vapor fraction.

In all the chemical lasers, optical methods are also used for measuring the uniformity of the gain medium, measurement of small signal gain by measuring amplification of a probe beam at their respective lasing wavelengths. In order to perform all of these measurements proper selection of photo detectors is required.

In order to detect and estimate laser pulse shape and make small signal gain measurements for HF-DF laser (2.7 - 3.4 µm), Indium Antimonide (InSb) is best suited because of its responsivity in this range. J10D series Indium Antimonide detectors or equivalents are high quality InSb photodiode providing excellent performance in 1 - 5.5 µm wavelength region. Typical technical specifications of an InSb photo detector for HF-DF laser is given in Table 2.11 which can be fulfilled by J10D series from M/s Judson Technologies.

Table 2.11. Technical specifications of InSb detector for HF-DF laser.

Parameters	Range
Active area (mm)	1
Peak responsivity (A/W)	3
Detectivity D^* @peak 1 kHz (cmHz$^{1/2}$W^{-1})	1×10^{11}
Back ground current (µA)	7
NEP @ peak 1 kHz (Pw/Hz$^{1/2}$)	6

Mercury Cadmium Telluride (MCT) is suitable for detection of 10.6 µm wavelength of CO_2 GDL for pulse shape, duration and small signal gain measurements. A typical technical specification of a MCT detector for CO_2 GDL is given in Table 2.12 which can be fulfilled by J15 series of M/s Judson Technologies or equivalent model.

Table 2.12. Typical technical specifications of MCT detector for CO_2 GDL.

Parameters	Range
Active area (mm)	1
Peak responsivity (V/W)	2
Peak wavelength (µm)	10.6
Temperature (^{0}C)	-65
Min Detectivity D^* @peak 10 kHz (cmHz$^{1/2}$W^{-1})	1×10^8

In COIL, the concentration of iodine molecules is measured using absorption spectroscopic method. Since iodine molecule has peak absorption at ~499 nm, thus, silicon photo diode is selected to detect this wavelength because silicon detector has significant responsivity at this wavelength. In addition chlorine utilization is measured at 330 nm which requires silicon detector. For laser power pulse detection (1315 nm) and singlet oxygen yield measurement (1270 nm) of COIL, Ge photodiode may be used. Table 2.13 shows typical technical specifications of Silicon and Germanium detector for COIL. Germanium photo detector from M/s Judson Technologies, USA J16 series and silicon photo detector from RS stock no. 303-674 or equivalent from M/s Hamamatsu, M/s Fairchild, M/s Optek etc. may be used for COIL application.

Table 2.13. Typical technical specifications of silicon and germanium detector for COIL.

Parameters	Silicon Detector for Iodine concentration/ Chlorine utilization measurement	Germanium Detector for Singlet oxygen Yield/ COIL pulse shape measurement
Spectral range (nm)	300-600	1100-1500
Active area (mm^2)	75-100	75-100
Typical responsivity (A/W)	0.5	0.5
Ambient temperature	-55 to 70 °C	-55 to 60 °C

References

[1] Olfa Kanoun and Hans-Rolf Trankler, Sensor Technology Advances and Future Trends, *IEEE Transactions on Instrumentation and Measurement*, Vol. 53, 6, Dec. 2004, p 1497.
[2] Khajan, Alexander D., Transducers and their elements, Eaglewood Cliffs, *Prentice Hall Publishing*, NJ, 1994.
[3] Allocca, John A., and Allen Stuart, Transducers Theory & Applications, *Reston Publishing, Inc.,* Reston, VA, 1984.
[4] Hauptmann P., Sensors-Principles and Applications, *Prentice Hall*, 1991.
[5] S. C. Mukhopadhyay, G. S. Gupta, Smart sensors and Sensing Technology, *Springer-Verlag, Berlin-Heidelberg*, 2008.

[6] Baker, H. D., E. A. Ryder and N. H. Baker, Temperature Measurement in Engineering, Vols. 1 and 2, *Wiley,* New York, 1953, 1961.

[7] American society for Testing and Materials: Manual on the use of Thermocouples in temperature measurement, *ASTM STP 470A*, March 1974.

[8] Benedict, R. P., and R. J. Russo, Calibration and Application Techniques for Platinum Resistance Thermometers, *Journal of Basic Engineering,* June 1972, p. 381.

[9] Becker, J. A., C. B. Green, and G. L. Pearson, Properties and Uses of Thermistors, *Trans. Am. Inst. Electr. Eng.,* Vol. 65, Nov 1946, p. 711.

[10] Travis, B, Temperature management ICs combat system meltdown, *EDN*, p. 38, 1996, August 15.

[11] Sachse, H., Semiconducting temperature sensors and their applications, *Willey*, New York, 1975.

[12] Franz Mayinger (Editor), Optical measurements: Techniques and Applications, *Springer-Verlag Berlin Heidelberg,* 1994.

[13] Brace, W. F., Effect of Pressure on Electrical-Resistance Strain Gauges, *Exp. Mech.,* Vol. 4, No. 7, 1964, p. 212.

[14] Ajluni, C., Pressure sensors strive to stay on top, *Electronics Design,* Oct. 3, 1994, p 67.

[15] Design considerations for Diaphragm Pressure Transducers, Technical note 105, *Measurement Group, Inc.,* 1982.

[16] Joost C. Lotters, Woulter Olthuis, Peter H. Veltink and Piet Bergveld, A sensitive differential capacitance to voltage converter for sensor applications, *IEEE Transactions on Instrumentation and Measurement,* Vol. 48, 1, Feb. 1999, p 89.

[17] Heerens W. C., Application of capacitance techniques in sensor design, *Phys. E., Sci. Instrum.,* Vol. 19, 1986, p. 897.

[18] Hanneborg A., Hansen T. E., Ohlckers P. A., Carlson E., Dohl B. and Holwech O., An integrated capacitive pressure sensor with frequency modulated output, *Sensors and Actuators,* 9, 1986, p 345.

[19] Strain gauge selection criteria, procedures and recommendations, Technical note 505, *Measurement Group, Inc.,* 1976, p. 1.

[20] Sagiyama, S., Takigawa M. and Igarashi I., Piezoresistive pressure sensor with both voltage and frequency output, *Sensors and Actuators,* 4, 1983, p. 113.

[21] http://www.metran.ru

[22] Thomas E. Kissell, Industrial electronics: Applications for programmable controllers, instrumentation and process control, and electrical machines and motor controls, *Prentice Hall, Inc.,* 2000.

[23] Shapiro A. H., The Dynamics and Thermodynamics of Compressible Fluid Flow, Vol. I, *Ronald Press,* New York., 1953.

[24] S. M. Yahya, Fundamentals of Compressible Flow, 2nd Edition, *New Age International (P) Ltd,*1998.

[25] D. L. Carroll, Overview of high energy lasers: Past, Present and Future, *AIAA* paper, 2011, pp, 2011-3102

[26] I. Rego, K. N. Sato, Y. Miyoshi, T. Ando, K. Goto, M. Sakamoto, S. Kawasaki, Studies on the characteristics of the gas dynamics laser with low CO_2-concentration medium by a diaphragmless shock tube, in *Proceedings of the 34th EPS Conference on Plasma Physics*, 31F, 2007, 0-4-036.

[27] G. Chizinsky, Recent advances in mass flow control, *Solid State Technology*, p 85, Sept 1994.

[28] Frank M. White, Fluid Mechanics, *McGraw-Hill*, New York, 1979.

[29] Arkilic E. B., Schmidt M. A., and Breuer K. S., Flow measurements near atmospheric pressures, *Experiments in Fluids*, 25, 1998, p. 37.

[30] Esashi M., Eoh S., Matsuo T. and Choi S., The Fabrication of integrated mass flow controllers, in *Proceedings of the Transducers '87*, 1987, p. 830.

[31] Victor L. Streeter and E. Benjamin Wylie, Fluid Mechanics, Seventh Edition, *McGraw-Hill Book Company*, 1979.

[32] Shapiro A. H., The Dynamics and Thermodynamics of Compressible Fluid Flow, Vol. I, *Ronald Press*, New York., 1953.

[33] Mainuddin, M. T. Beg, Moinuddin, R. K. Tyagi, R. Rajesh, Gaurav Singhal and A. L. Dawar, Real time gas flow control and analysis for high power infrared gas lasers, *International Journal of Infrared and Millimeter Waves*, Vol. No. 1, Jan. 2005, p. 91.

[34] James W. Dally, Rilley W. R. and McConnell K. G., Instrumentation for engineering measurements, Chapter 5, *John Wiley & Sons, Inc.,* 1993.

[35] Govind P. Agrawal, Fiber-optic communication systems, Chapter IV, *John Wiley & Sons, Inc.,* 1992.

[36] Kasper B. L., Optical fiber transmission II, S. E. Miller and I. P. Kaminow, Eds., *Academic Press*, Chapter 18, 1988.

[37] M. Dombrowski and P. Wilson, *SPIE Proc.,* 3753, January 2000, 100.

Chapter 3

Diagnostic Techniques

In the earlier chapters we have illustrated the significance and methods of determining flow parameters of various gas effluents viz. flow rate of individual gas constituents, pressure, and temperature, as they are very crucial for laser operation.

However, these alone are not sufficient for characterizing and optimizing the output of these high chemical gas lasers. Apart from basic parameter sensing and measurement, these advanced chemical laser systems also require customized techniques for evaluating the laser performance. These take inputs from several direct sensors for evaluation of physical parameters *(derived parameters)*. This is essential since many of the questions regarding their operation are still unanswered requiring sensitive diagnostics, which may provide critical insights into these scientifically interesting laser systems.

This chapter enunciates the various diagnostics techniques applicable to chemical laser systems. The diagnostics techniques have been grouped on the basis of the generic parametric measurement or characterization being carried out.

1)   Specie concentration measurement
   - a.   Optical emission
   - b.   Optical absorption
   - c.   Diode laser based absorption spectroscopy
   - d.   Cavity Ring Down Spectroscopy
   - e.   Raman Spectroscopy
2)   Cavity Medium Characterization
   - a.   Mach number
     - i.   Pitot static tube method
     - ii.   Laser Doppler Velocimetery
     - iii.   Voigt Profile method
   - b.   Small Signal Gain measurement
     - i.   Probe beam method
     - ii.   Voigt Profile method

    c.   Medium Homogeneity
        i.   Gain mapping
        ii.   Optical Interferometery
        iii.   Laser Induced Fluorescence (LIF)/ Planar-LIF (PLIF)
3)   Laser Power and pulse measurement
    a.   Calorimetric method
    b.   Pulse shape and duration

Table (3.1) presents the applicable diagnostics techniques for measurement of numerous system parameters in different chemical lasers. The applicability for various lasers has been mentioned on the basis of their usage in the available literature.

However, it would be prudent to mention that diagnostic techniques being discussed here are by no means comprehensive since extensive research is being carried out on these systems resulting in regular development of newer diagnostics setups.

3.1 Specie Concentration Measurement

Species concentration measurements [1] are critical for operation of chemical lasers systems. The measurements allow ascertaining whether the systems are operating at design flow rates for various species and may be utilized as a tool for troubleshooting as well. Both excess and lack of specific species lead to degradation of desired laser power levels.

In general on line estimation of specie concentration of the lasing and pumping mediums depending on the laser, using optical/non-optical methods is imperative. Since optical methods are by far non intrusive, hence are the preferred mode of measurements in flowing medium chemical gas lasers.

Further, since some of these lasers involve intensive reactive processes which may be multiphase as well, hence determination of concentration of chemical species is not limited to lasing /pumping but may also involve intermediate reactants/ products, water vapor etc.

Table 3.1. Applicable diagnostics for various parameters
for different chemical lasers.

Measured Parameter	Diagnostics	Gas chemical laser
Iodine concentration (I_2)	a) Optical absorption b) Cavity ring down spectroscopy (CRDS)	COIL
Chlorine Utilization (Cl_2)	a) Optical absorption b) Cavity ring down spectroscopy (CRDS)	COIL
Sulphur Hexafluoride (SF_6)	Optical absorption	HF/DF
Singlet Oxygen Yield (Y_{O2*})	a) Optical emission b) Diode laser based diagnostics c) Raman spectroscopy	COIL
Water vapor concentration (H_2O)	a) Optical emission b) Diode laser based diagnostics	COIL
Mach number	a) Pitot static tube b) Lased Doppler Velocimetery (LDV) c) Voigt Profile method	COIL, GDL, HF/DF GDL COIL
Small signal gain	a) Probe beam diagnostics b) Voigt Profile method	COIL, CO_2 GDL, HF/DF COIL
Homogeneity	a) Gain mapping b) Optical Interferometery c) Laser Induced Fluorescence (LIF) d) Planar laser Induced Fluorescence (PLIF)	CO_2 GDL, HF/DF COIL COIL
Laser power	Calorimeter	COIL, CO_2 GDL, HF/DF
Laser pulse shape	Optical emission	COIL, CO_2 GDL, HF/DF

3.1.1 Optical Emission

Light emission by excited atoms or molecules collected via an optical setup mainly comprising of an interference filter and a suitable photo detector compatible with the emission wavelength form the basis of optical emission diagnostics.

Emission is a more generic term, however in chemical lasers the emission occurs from energized molecules generated by chemical mechanism, which is mostly termed as *chemiluminescence* [1].

Fig. 3.1 shows the typical schematic for an optical emission diagnostics. The emission from the specie to be measured is collimated using a lens followed by a filter for eliminating photons from unwanted wavelengths. The emitted photons are the allowed to fall on to the detector sensitive to the measured wavelength. The detector signal is then fed to the data acquisition system for further processing. Also, simultaneously, pressure and temperature of the medium is also measured for correlating with specie concentration.

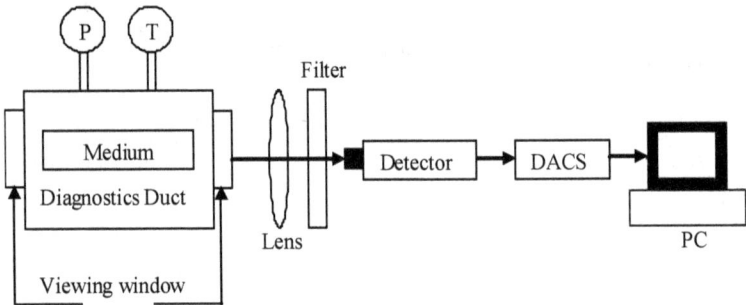

Fig. 3.1. Schematic for the optical emission diagnostics.

The governing equation for emission signal detected by the photo detector is given as

$$S_\lambda \quad \alpha \quad [C_i]\eta_\lambda T_\lambda d\Omega \qquad (3.1)$$

The photodiode signal (S_λ) is proportional to the measured specie concentration, sensitivity and collection efficiency. Where, C_i is the concentration of the measured species (in all subsequent equations square brackets have been dropped), η_λ is the detector collection efficiency, T_λ is the resultant transmission and $d\Omega$ is the solid angle. The subscript λ is the emitting wavelength of the specie. The above relation can also be written as:

$$S_\lambda = AC_i ,$$ (3.2)

where A is the calibration factor of the entire detection system. The concentration may again be written as function of pressure and temperature of the medium to yield the following relation,

$$S_\lambda = A\frac{P_i}{kT}$$ (3.3)

k is the Boltzmann constant, P_i is the partial pressure of the measured specie, T is the gas temperature.

The application of the above described diagnostics setup is illustrated below by taking two examples from COIL viz., i) singlet oxygen yield measurement, ii) water vapor fraction measurement.

3.1.1.1 Singlet Oxygen Yield

In case of chemical oxygen iodine laser, chemical generation of singlet oxygen is a complex phenomenon involving a two-phase interaction between chlorine gas and BHP liquid. As has also been discussed in Chapter -1, the generated singlet oxygen may also quench to generate ground state oxygen. Hence, the outflow will comprise of unutilized chlorine (measurement discussed later in *optical absorption*), ground state oxygen, singlet oxygen apart from buffer nitrogen introduced to reduce transport losses.

Thus, parameters in terms of generator pressure, relative momentum ratio of liquid reagent and gas, BHP temperature, geometry of flow extraction for the generation of the singlet oxygen-pumping medium needs to be well controlled. These parameters greatly affect the efficiency of the generator i.e. singlet oxygen yield.

Hence, it is important to accurately quantify the actual amount of singlet oxygen generated after the involved losses to be able to characterize the generator for achieving its optimal operation. Thus, there is a need to develop dedicated measurement system for singlet oxygen using non-intrusive type optical technique.

The meta-stable singlet oxygen molecules O_2 ($^1\Delta$) emit at 1.27 μm [2]. The emitted photons may be collected using liquid nitrogen cooled Germanium photodiode (M/s EG & G Judson, Model No. JD16-M204) of reasonable diameter (say 5mm) diameter through a lens (say 25 mm size) fixed to the diagnostic cell. An interference filter with $\lambda = 1270 \pm 5$ nm is introduced between the lens and the photodiode for reducing the unwanted photons reaching the photo detector.

The diagnostic duct or plenum placed at the generator exit generally has provision for two optical windows of suitable size (say one-inch windows) for O_2 ($^1\Delta$) emission measurements. Also, provisions for plenum pressure and temperature measurement are also provided. Further, provisions for taking a bleed, in order to probe the gas sample for chlorine utilization measurement may also be included in the same diagnostic duct (details of optical set up discussed later under optical absorption). Fig. 3.2 shows the developed hardware for singlet oxygen measurement.

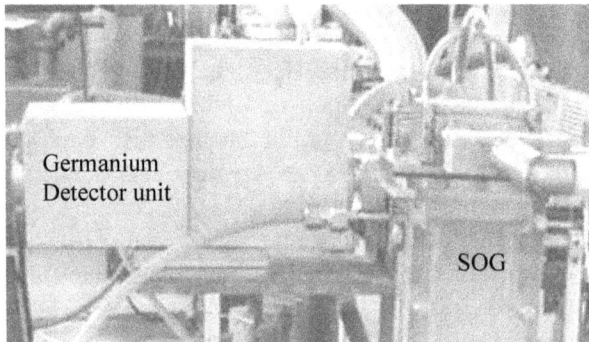

Fig. 3.2. Developed diagnostics set up for singlet oxygen.

Thus, from Eq.(3.3) the signal corresponding to emission for singlet oxygen molecules is given by,

$$S_\Delta \;\; = \;\; A \frac{P_\Delta}{kT}, \tag{3.4}$$

where P_Δ is the singlet oxygen partial pressure and T is the gas temperature. From the measurement of the singlet oxygen concentration, the yield can be estimated as:

$$Y_\Delta = \frac{[O_2(^1\Delta)]}{[O_2]_{total}} = \frac{C_\Delta}{C_{O2tot}} = \frac{P_\Delta}{P_0}, \qquad (3.5)$$

where the P_o is the total oxygen partial pressure in the diagnostic duct comprising of ground state and singlet delta oxygen. It can be calculated for the known values of chlorine M_{Cl2} and buffer gas flow rates M_{buffer}, cell pressure P and chlorine utilization U_{cl} (discussed later in *optical absorption* section) in accordance with the relation:

$$P_0 = \frac{PM_{Cl_2}U_{Cl}}{M_{Cl_2} + M_{buffer}} \qquad (3.6)$$

Eq. (3.5) can be rewritten as using Eq. (3.4) and Eq. (3.6),

$$Y_\Delta = kTS_\Delta \frac{(M_{Cl} + M_{buffer})}{APM_{Cl2}U_{Cl}} \qquad (3.7)$$

The calibration of system is generally carried out by the technique suggested by Zagidullin [3]. A mixture of chlorine and nitrogen (1:9 ratio) is fed to the SOG through chlorine supply line. The typical SOG pressure is maintained at 30 torr with gas velocity of nearly 15 ms^{-1}. The corresponding gas residence time is ~6.7 ms^{-1} in the reaction zone which is within the range of complete utilization of chlorine. Under these conditions, partial pressure of chlorine (3 torr) inside SOG is low enough to neglect the pooling losses and a yield of ~100 % is expected. Based on several experiments keeping constant SOG pressure and varying cell pressure, calibration factor 'A' may be determined.

In our experiments, consisting of twin SOG module comprising a total chlorine flow rate of 1500 mmols^{-1}, the determined value of 'A' is 0.33 Vtorr^{-1} and the observed singlet oxygen yield is ~64 % [4].

3.1.1.2 Water Vapor Concentration

In COIL operation, production of singlet oxygen is a two-phase reaction inevitably leading to production of water vapor. In COIL presence of water vapor leads to quenching of singlet oxygen and excited iodine molecules and atoms as well, hence water vapor fraction greater than a typical value of ~ 5 % is detrimental to laser functioning.

Thus, measurement of water vapor is essential for being able to determine the amount of water vapor present in the flow and taking necessary corrective action if required. The corrective action generally constitutes of adjusting the BHP concentration and its working temperature.

The water vapor concentration is measured using an optical emission methodology suggested by Spalek *et. al* [5]. The principle of measurement is based on the following reactions:

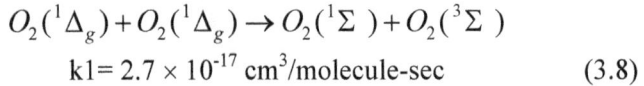

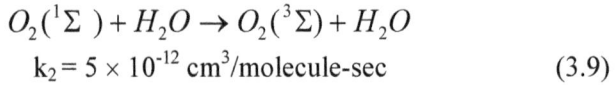

$$O_2(^1\Delta_g) + O_2(^1\Delta_g) \rightarrow O_2(^1\Sigma) + O_2(^3\Sigma)$$

$$k1 = 2.7 \times 10^{-17} \text{ cm}^3/\text{molecule-sec} \tag{3.8}$$

$$O_2(^1\Sigma) + H_2O \rightarrow O_2(^3\Sigma) + H_2O$$

$$k_2 = 5 \times 10^{-12} \text{ cm}^3/\text{molecule-sec} \tag{3.9}$$

The first reaction (3.8) indicates the pooling of two singlet delta oxygen molecules to form a singlet sigma molecule along with a one ground state oxygen, whereas, the second reaction depicts the de-excitation of the singlet sigma molecule into the ground state oxygen molecule on interacting with water molecule. Since the rate of the second reaction is several orders higher than the first reaction, the presence of the singlet sigma concentration in the flow directly reflects the presence of water vapor molecules. At any given point in time, the concentration of singlet sigma molecules is directly proportional to the square of the singlet delta molecules and inversely proportional to that of the water molecules, as given by Eq. (3.10),

$$\left[O_2\left(^1\Sigma\right)\right] = \frac{k_1}{k_2} \frac{\left[O_2\left(^1\Delta\right)\right]\left[O_2\left(^1\Delta\right)\right]}{\left[H_2O\right]} \tag{3.10}$$

The water vapor concentration can therefore be written as below:

$$C_{water} = \frac{k_1}{k_2} \frac{C_\Delta^2}{C_\Sigma}, \tag{3.11}$$

where C_Δ and C_Σ are the $O_2\,(^1\Delta)$ and $O_2\,(^1\Sigma)$ concentrations respectively and k_1 and k_2 are the rate constants of the corresponding reactions.

Thus, the diagnostics for water vapor concentration requires the measurements of emission signals from $O_2\,(^1\Delta)$ (1.27 μm) and $O_2\,(^1\Sigma)$ (0.762 μm). Therefore, in terms of the photo signals Eq. (3.12) can be written as:

$$C_{water} = C_{relative}\,\frac{S_\Delta^2}{S_\Sigma}, \qquad (3.12)$$

where S_Δ and S_Σ are the corresponding photo detector signals and $C_{relative}$ is the calibration factor and is determined using relations for water vapor saturation pressure [5, 6] given as follows,

$$P_w = 1.6x10^{11}e^{\left(\frac{-5294}{T}\right)}x_w\,\exp\left(\frac{-344.2}{\left(T\left(1+0.8053\dfrac{x_w}{x_p}\right)\right)^2}\right), \qquad (3.13)$$

where x_w and x_p are the molar fractions of water and peroxide and T are the BHP jet temperature. The diagnostic scheme that may be employed is similar to as shown in Fig. 3.1 with both optical windows being utilized one for measurement of $O_2\,(^1\Delta)$ signal and the other for measurement of $O_2\,(^1\Sigma)$. The photodiode need to be chosen corresponding to the compatible emission wavelength i.e. for $O_2\,(^1\Delta)$ (1.27 μm) Ge photodiode is used and for $O_2\,(^1\Sigma)$ (0.762 μm) Si-photodiode is employed.

The emphasis of the above discussion is primarily to bring out the fact that in order to design an effective diagnostics one has to be well aware of the theory of kinetics of generation of the specific species.

3.1.2 Optical Absorption

Optical absorption technique utilizes the property of the species to absorb light at specific wavelengths. It is a simple, noninvasive, in situ technique for obtaining information about gas phase species. From an absorption spectrum, quantitative absolute concentrations and absolute frequency- dependent cross sections may be extracted.

It basically employs a suitable light source, to generate probe beam, followed by a collimating lens and an interference filter. The beam having passed through the medium is then collected by a detector via a focusing lens. The schematic of the setup that may be employed is shown in Fig. 3.3.

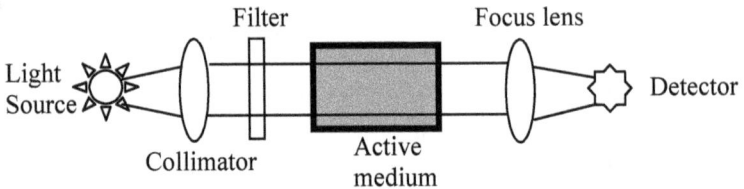

Fig. 3.3. Schematic for Optical absorption diagnostics.

The basic principle is that the beam is passed through the medium whose concentration/ density is to be measured. On the other side, the amount of light transmitted by the medium will be measured and the reduction is proportional to the number density of the medium. The phenomenon follows the Beer Lamberts law:

$$\frac{I_v}{I_o} = e^{(-\sigma_v nL)},$$
(3.14)

where

I_v is the transmitted light intensity at frequency v;

I_0 is the incident light intensity (intensity without the medium);

σ_v is the medium absorption cross – section at probe beam wavelength (m^2) (is estimated from the relation for molar extinction coefficient) ($\in$);

n is the species concentration (molecules per cm^3);

L is the path length (m);

Each species has its own absorption spectrum and hence the light source used should be selected accordingly. Further, the absorption cross-section for a medium will be different for different wavelength. Therefore, it is always preferable to use a monochromatic light source for this purpose, the wavelength of which is near or equal to the peak absorption wavelength of the medium. On the other hand, if one uses a broadband light then one has to use the normalized (average) absorption cross-section for the estimation. Practically, to use a monochromatic light source for producing the probe beam corresponding to the peak absorption wavelength of the medium is a difficult task. Therefore, one can use a broad band light source in combination with a narrow band interference filter corresponding to the wavelength of interest as shown in Fig. 3.3.

If the medium contains only the species of interest and the active volume is a sealed one then the above relation (3.14) may be directly applied provided the absorption cross-section is known. In case, the medium is a combination of more than one species and the active volume is a flowing medium then the problem has to be handled in a different manner as discussed below.

In the latter case, one has to know the flow rate of the other species and the total pressure of the medium in the active volume so that the flow rate of the species of interest can be determined as follows.

$$n = \frac{P_i}{kT} \tag{3.15}$$

The above relation yields the number density of the species of interest. Where, k is the Boltzmann Constant ($=1.38 \times 10^{-23}$J/K), T is the temperature of the medium; P_i is the specie partial pressure. Therefore, Beer's law can also be written as:

$$P_i = \left\{ \frac{kT}{\sigma L} \right\} \ln \left(\frac{I_o}{I_v} \right) \tag{3.16}$$

The corresponding specie flow rate (M_{12}) into the system may then be estimated using the following relation (By Dalton's law),

$$M_i = \frac{P_i}{P - P_i} M_c, \tag{3.17}$$

where P is the total cell pressure of the active volume and M_c is the carrier gas (N$_2$) molar flow rate generally employed in chemical gas lasers.

The use of above diagnostics is discussed by considering typical examples of measuring Chorine, Sulphur Hexafluoride (SF$_6$), and Iodine.

3.1.2.1 Chlorine Utilization (Cl$_2$) Concentration

It has already been discussed earlier that chlorine is one of the most essential gas reagents employed in COIL required for generation of the singlet oxygen-pumping medium. Chlorine molecules have peak absorption at about 330 nm [7] having absorption cross-section (σ) of 2.75×10^{-19} cm^2.

A portion of the singlet oxygen generator (SOG) flow exit medium is passed through optical cell, which is continuously pumped out during operation for maintaining the identical conditions as in the main duct. The cell has suitable viewing windows along with provisions for measurement of pressure and temperature carried out using suitable sensors as discussed earlier.

An Ultra Violet mercury lamp is used as a light source along with an interference filter with 330 ± 5 nm. The collimated light is passed through the cell and detected using silicon photodiode (RS component stock no. 303-674). The flow rate of exit chlorine ($M_{Cl2})_{exit}$ can be estimated using the relation (3.14) to (3.17) discussed earlier for multiple species case.

The chlorine input flow rate ($M_{Cl2})_{input}$ is known using the principle of orifice under choke flow condition also discussed earlier in Chapter 2. Therefore, chlorine utilization has been estimated using the relation (3.18).

$$U_{Cl} = 1 - \frac{(M_{Cl2})_{exit}}{(M_{Cl2})_{input}} \qquad (3.18)$$

Fig. 3.4 shows the photograph of chlorine utilization measurement system used in COIL application.

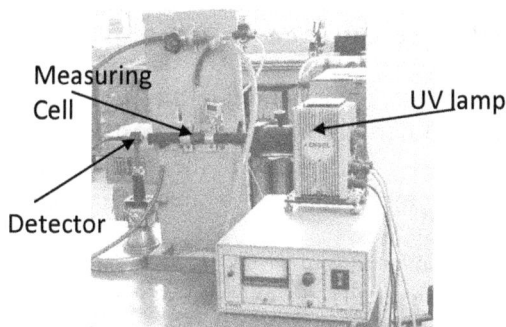

Fig. 3.4. Developed system for chlorine utilization.

The experiments have been conducted for large variation in chlorine flow rate conditions. It has been observed that the system is capable of measuring unutilized chlorine traces of the order of ~0.8 mmols^{-1}. The detailed experimentation on the COIL source used shows a typical chlorine utilization of nearly 94 % [8].

3.1.2.2 Sulphur Hexafluoride (SF$_6$) Concentration

As explained earlier, SF$_6$ is the source of fluorine atoms essential for generation of HF/DF laser. SF$_6$ is chosen as a fluorine carrier since it is non-toxic, non-corrosive gas that can be handled safely in large quantity.

Since supplied SF$_6$ may have traces of water vapor or oxygen, hence it is appropriate to choose a source wavelength of < 140 nm as water vapor has maximum absorption at 165 nm and oxygen at 142 nm. The highest absorption coefficient ($n\sigma$, units of m^{-1}) is 10^4 m^{-1} [9] and occurs for a wavelength of 115 nm.

The SF$_6$ diluted with nitrogen is passed through the diagnostics cell. The cell has suitable viewing windows along with provisions for measurement of pressure and temperature carried out using suitable sensors as discussed earlier.

A hydrogen argon lamp equipped with MgF$_2$ windows may be used as the light source coupled with a suitable interference filter. The collimated light passed through the detection cell is detected using Hamamatsu model R 268 photomultiplier tubes. The same set of relations from (3.14) to (3.17) can be used for estimation of the flow rates.

3.1.2.3 Iodine (I₂) Concentration

In case of COIL, the iodine is carried to the laser head from the iodine evaporator using hot nitrogen gas. The diagnostic system is located just after the exit of gas mixture from the evaporator. Therefore, the medium is again a combination of more than one species.

The gain and the power output of COIL are extremely sensitive to the iodine flow rates. Typically, in case of subsonic systems, the power decreases significantly when the iodine concentration changes even by 5 %, however, for supersonic COIL [10, 11], the concentration range is much broader.

Although, the operating ranges of iodine concentration for typical COIL operation have been established, however, each individual system is unique in terms of its nozzle geometry, optical resonator configuration, iodine injection parameters etc. Therefore, different workers have observed various values of optimal Iodine concentration for best COIL performance. Thus, the iodine concentration or flow rate measurement is an important COIL diagnostics for precisely determining the real time iodine flow during the run duration.

Iodine molecules exhibit peak absorption at close to 490 nm with an absorption coefficient of 2.1×10^{-18} cm^2 [12].

The diagnostics set up consists of a probe cell with suitable viewing windows along with provisions for measurement of pressure and temperature carried out using suitable sensors as discussed earlier.

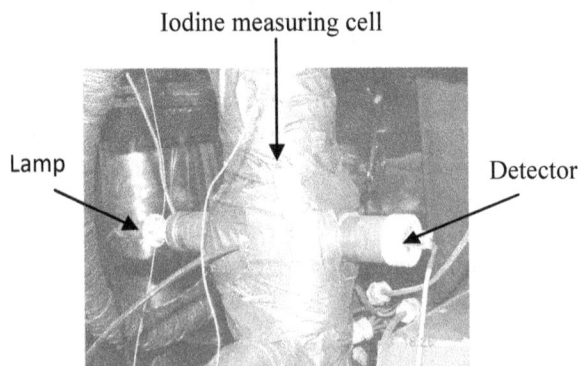

Fig. 3.5. Photograph of Iodine concentration measuring system.

A tungsten halogen lamp of suitable power (say 50 W), having a broad wavelength spectrum, along with an interference filter of wavelength band 490 ± 5 nm may be used as the light source. The beam is collimated using a collimating lens assembly before passing through the cell. The typical hardware developed for a COIL system [13, 14], is shown in Fig. 3.5. The transmitted light is focused on to a silicon photodiode (RS component stock no. 303-674) along with signal-conditioning module for the detection purposes. The amplifier output is fed to the PC through an analog input of the PCI based data acquisition card (PCI-1716) for acquisition, storage and analysis. Alternatively, an Ar ion laser emitting at 488 nm may also be used as source for iodine detection, in which case collimation optics will not be required. The same set of relations from (3.14) to (3.17) may be used to calculate iodine mass flow rate.

3.1.3 Diode Laser based Absorption Spectroscopy

Use of diode laser based diagnostics is becoming increasingly common in case of COIL gas lasers. The principles of optical absorption have already been outlined in the earlier section. In this session, we will just enunciate the generic methodology being used for determination of specie concentration.

Lately, Physical Sciences Inc. [15] developed several of diode based diagnostics setup, which may be employed not only for COIL but also for ECOIL systems. As far as specie concentration is concerned, the three most critical species to be measured are chlorine utilization, water vapor fraction and singlet oxygen yield.

Chlorine utilization is found by the number density of unreacted Cl_2 in the cell as has been discussed earlier.

The water vapor diagnostics is based on a diode laser, which scans an individual rotovibration line of water v_1+v_2 band in 1.39-μm region. The laser beam is split into signal and reference beams. The system can supposedly measure water molecule densities of the order of 10^{15} cm^{-3}. The prime error occurs from a level of arbitrariness that in integration of profile borders in the provided software. The expected error in water vapor fraction is no greater than ± 0.01 %.

The singlet oxygen diagnostic system measures oxygen ground state O_2 ($^3\Sigma$) density. It is based on diode laser that scans an R5R5 line of the O_2

$(^3\Sigma) \rightarrow O_2 (^1\Sigma)$ electronic – rotation transition. The wavelength of the transition measured by the manufacturer of the diagnostic systems (Physical Science Inc.) is 761.2 nm. The beam is split into reference beam and signal. Typically, since the absorption signal is weak, the signal is passed 21 times through a diagnostic cell (typically 5 cm wide) in a multi pass Herriott cell configuration. The beams are compared by a balanced ratiometric detector, which amplifies the logarithm of the ratio between the reference and the signal. The diagnostic is calibrated at the start and end of each run by measuring the area under the absorption line for medical grade oxygen (99 % purity) at expected partial pressures of 1 to 10 torr. The sources of error could be due to non-linearity, aerosol droplets on the windows, and changes in gas temperature. Typically, the error in yield measurement is expected to be of the order of ± 0.05 %.

Singlet oxygen yield is inferred from the fact that, whatever is remaining apart from chlorine, water vapor or ground state oxygen is singlet oxygen i.e. $O_2 (^1\Delta)$. This assumption is justified by the fast quenching of $O_2 (^1\Sigma)$ with water vapor as has already been discussed earlier.

$$Y_\Delta = 1 - \frac{P_{og}}{P_0 - P_{Cl} - P_w}, \qquad (3.19)$$

where Y_Δ is the singlet oxygen yield, P_{og} is the partial pressure of ground state oxygen, P_{Cl} is the partial pressure of chlorine and P_w is the partial pressure of water vapor.

3.1.4 Raman Spectroscopy

Raman spectroscopy is a powerful experimental tool to identify and quantify the species present in the sample. Recently it is being developed for diagnostics of singlet oxygen and ground state oxygen in COIL system. It has several advantages over the optical emission method discussed earlier since it allows one to monitor both singlet and ground state oxygen simultaneously in the same measurement volume. The common experimental problems such as occurrence of dirty windows, fluctuations in source power, aerosol scattering can be ratioed out. Thus, the present session discussed its use in reference to COIL.

However, the application of Raman spectroscopy in COIL is not fairly simple and is full of challenges. The major problem areas are due to high-speed flow of mixture gases like singlet oxygen, oxygen, residual chlorine and nitrogen at relatively low pressure in the measurement cell. In addition, Raman scattering cross-section of gases is extremely small ($\sim 10^{-30}$ /cm^2). These facts make the Raman signal intensity very low and it becomes difficult to record the Raman signal even with the state of the art CCD detectors. The schematic of Raman diagnostics [16] is shown in Fig. 3.6.

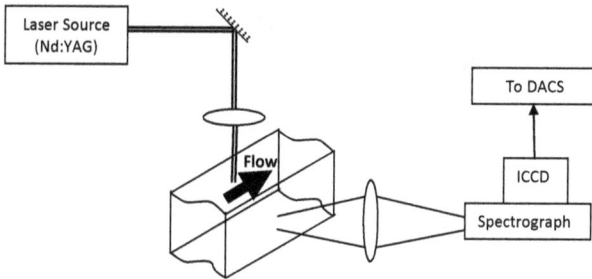

Fig. 3.6. Schematic of Raman scattering diagnostics.

In the Raman spectroscopy, a pump laser such as frequency doubled Nd:YAG laser (Repetition rate $\sim$ 10 Hz and pulse duration $\sim$ 20 ns) emitting at 532 nm with pulse energy of nearly 250 mJ/ pulse may be employed. It is focused at the center of the gas flow and Raman scattering is collected perpendicular to the direction of the flow by a combination of lenses and fiber. The fiber is connected to the spectrograph with CCD detector, which records the spectra. The intensity of Raman spectrum is decided by scattered molecule structures. Typically, stokes line for ground state oxygen is at 580 nm and for singlet oxygen occurs at 577 nm. The intensity of Raman signal is directly proportional to the concentration of molecule in sample. Thus, the concentration of singlet oxygen and ground state oxygen relative to nitrogen concentration may be written as:

$$A_1 = \frac{\left[\dfrac{I_{O_2(1\Delta)}}{\sigma_{O_2(1\Delta)}} \right]}{\left[\dfrac{I_{N_2}}{\sigma_{N_2}} \right]} \tag{3.20}$$

$$A_2 = \frac{\left[\dfrac{I_{O_2(3\Sigma)}}{\sigma_{O_2(3\Sigma)}}\right]}{\left[\dfrac{I_{N_2}}{\sigma_{N_2}}\right]}, \qquad (3.21)$$

where A_1 and A_2 are the concentration ratio of the $O_2(^1\Delta)$ and $O_2(^3\Sigma)$ respectively, to N_2 at the exit of the SOG; I is the area under the respective peaks of the Raman spectrum of $O_2(^1\Delta)$, $O_2(^3\Sigma)$, and N_2; and σ is the Raman cross section. Hence, the concentration ratio of the total oxygen to nitrogen at the exit of the SOG can be obtained by using the expression,

$$\frac{[O_2]}{[N_2]} = A_1 + A_2 \qquad (3.22)$$

Thus, the yield of singlet oxygen can be calculated using

$$Y = \frac{\left[O_2\left(^1\Delta\right)\right]}{\left[O_2\left(^1\Delta\right) + O_2\left(^3\Sigma\right)\right]} \qquad (3.23)$$

Inserting the values of molecular concentrations in terms of Raman scattering cross-sections and intensities, we obtain the singlet oxygen yield as

$$Y = \frac{\left[I_{O_2(1\Delta)}\right]}{\left[I_{O_2(1\Delta)} + I_{O_2(3\Sigma)}\left(\dfrac{\sigma_{O_2(1\Delta)}}{\sigma_{O_2(3\Sigma)}}\right)\right]} \qquad (3.24)$$

The above equation can be directly used to calculate the singlet oxygen concentration from SOG.

Similar methodology may be used for targeted detection of certain specific species in other chemical gas lasers as well using a similar methodology.

3.1.5 Cavity Ring Down Spectroscopy (CRDS) for Trace Detection of Gases

Optical absorption based concentration measurement methods are suitable for measurement of sufficiently large flow rates or concentration of gas species. However, in case of detection of certain species or gases existing in trace amounts or having a very low absorption cross-section, direct absorption spectroscopy is not suitable as it suffers from low sensitivity. This low sensitivity results from the fact that a small light attenuation has to be measured on top of a large background signal that in turn is proportional to the intensity of the light source. As a result, pulsed lasers, which cover a broad wavelength region (from the UV to the IR), but exhibit large pulse-to-pulse intensity fluctuations, are not well suited for these studies. Often, however, these techniques cannot be used in certain environments and cannot detect or measure low concentrations of a substance.

A simple extension of Laser absorption spectroscopy is Cavity Ring–Down Spectroscopy (CRDS) [17, 18]. CRDS was conceived in 1988 by two researchers, Anthony O'Keefe and David Deacon of Deacon Research at Palo Alto, US. Cavity ring-down spectroscopy (CRDS), also termed as cavity ring-down laser absorption spectroscopy (CRLAS), is a laser absorption technique which takes advantage of the improved analytical sensitivity that is possible when making absorption measurements using extremely long path lengths. Effective path lengths of up to a few kilometers can be achieved by using a high-finesse optical cavity consisting of two high-reflectivity mirrors (R>99.9 %) to literally "bounce" a pulsed laser back and forth through the absorbing species on the order of about 10,000 times. The back and forth movement of the pulse is called ringing down. The schematic is shown in Fig. 3.7.

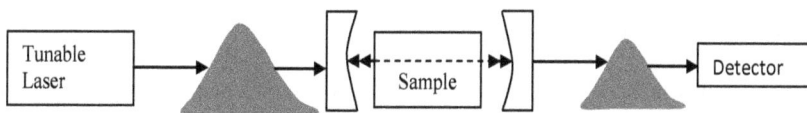

Fig. 3.7. Schematic of Cavity Ring down spectroscopy.

The light that is transmitted through the exit mirror on each pass is measured vs. time. The leaked signal is detected through fast detector and oscilloscope. Under certain conditions, the resulting signal will decay exponentially with time. The decay time will be dependant upon the reflectivity of the two mirrors, the distance between the two mirrors, the speed of light, and the molecular absorption coefficient of any absorbing species in the cavity. The exponential decay of signal directly scales with the absorption coefficient of the sample, which may be expressed as:

$$\alpha(\lambda) = \frac{1}{c}\left(\frac{1}{\tau(\lambda)} - \frac{1}{\tau_0}\right),$$
(3.25)

where τ_0 is the decay time without sample and $\tau(\lambda)$ is the decay time with sample at wavelength (λ). Thus in CRDS, by measuring the two decay times, absorption of the sample may be easily detected. The decay rate of a light pulse circulating inside a ring down cavity depends on the reflection efficiency of the mirrors and the attenuation by gases and aerosol particles present between the mirrors. Variations in the ringdown time measured at a single wavelength can then be linked to variations in the concentration of the target absorber.

CRDS can now be performed using semiconductor diode lasers, offering cost, stability, power and size advantages over pulsed sources. Not only that, the narrow line-width of the laser means it offers yet further sensitivity advantages- although not without some extra cost. Diode lasers may not have the frequency range of OPO- based lasers, but their advantages are attracting commercial enterprise to the technology.

CRDS is enormously sensitive. It picks up absorptions that are one-millionth of the strength of those that can be detected Fourier-transform infrared spectroscopy. It is also unaffected by laser noise. In this technique, 1 m cavity length with mirrors 99.95 % reflection can safely measure concentrations up to 100 ppt.

The technique has already found wide-ranging research applications, such as detecting trace amounts of volatile organic molecules, impurities and other chemicals species in plasmas, flames and discharges. The technique requires the observation of a finite amount of light decaying over a short time. This is easier using a tunable laser. A system based on optical parametric oscillation (OPD) is the natural choice, but makes the apparatus unwieldy, power-hungry, expensive and anything but portable. The more recent finding that continuous wave lasers can be used has taken the technique into another league-at least for certain applications.

This kind of diagnostics has been utilized for measurement of species in case of COIL such as absolute density of O_2 ($X^3\Sigma_g$) is probed by measuring the O_2 ($b^1\Sigma_g$) [v= 0] $\leftarrow$ O_2 ($X^3\Sigma_g$) [v = 0] transition near 760 nm or trace iodine detection at 490 nm or chlorine detection at 330 nm in small-scale systems. With the developmental focus shifting to ECOIL systems, this technique is bound to find applications for detection of numerous trace species produced in discharge production of singlet oxygen.

It would be appropriate to mention that in COIL applications researchers have used variation of CRDS termed as Off –axis CRDS or Off axis Integrated Cavity Output Spectroscopy (Off axis - ICOS) [19].

By employing an off-axis alignment scheme, the light does not interfere strongly with itself in the cavity, allowing very small changes in intensity to be measured while retaining extraordinarily long path lengths. Furthermore, these systems offset the problem of high sensitivity to alignment in case of CRDS, resulting in a large decrease in transmitted power even with slightest beam steering. Off-axis ICOS is robust and self calibrating since path length depends on losses and exact beam alignment allowing it to take high degree of mechanical vibration that may occur in case of field equipment. Moreover, the technique becomes self-calibrating by rapidly switching the laser off and measuring the decay of light out of the cavity, similar to the well-established technique of CRDS, but without dithering either the laser or cavity to hit a specific resonance or the "modematching" the beam.

3.2 Cavity Medium Characterization

In chemical gas lasers, it is imperative that one is able to characterize the critical medium parameters inside the cavity. Mach number is one such parameter which influences the laser power and needs to be estimated accurately. Also small signal gain is a representation of the extent of population inversion in the cavity and it is essential that we are able to accurately quantify the gain inside the cavity. Similarly, chemical laser systems involve mixing of multiple gas streams and a method of determining and being able to quantify a supposedly non-tangible parameter such as extent of mixing or medium homogeneity needs to be worked out. The present session discusses the measurement of these prime parameters of Mach number, small signal gain and medium homogeneity.

Hence, the session below discusses foremost the methods for determination of small signal gain followed by Mach number and medium homogeneity.

3.2.1 Small Signal Gain (SSG) Measurement

In case of gas lasers, small signal gain (SSG) is one of the prime influences on the amount of power extracted from the system. It basically quantifies the extent of population inversion achieved inside the cavity of a gas laser. In all the gas lasers, it is important to know the quantitatively the small signal gain of the lasing species. In case of CO_2 GDL, gain medium is CO_2 molecule and lasing occurs at 10.6 μm whereas in HF/DF laser gain medium is HF or DF with lasing at 2.7 or 3.7 μm and COIL lasers at 1.315 μm. One of the simplest ways for measurement of small signal gain is to use probe beam method and the other is the Voigt Profile method, which is primarily the extension of the former. Both of which are taken up in the succeeding sections.

3.2.1.1 Probe Beam Method

Probe beam measurement technique [20] has been extensively used for estimation of the small signal gain in gas lasers. The technique is based on Beer-Lambert's law in which the change (basically amplification) in beam intensity is governed is proportional to the number density of the excited atoms or the population inversion. The present discussion is taken up with reference to COIL laser but is easily extendable for the other two chemical gas lasers i.e. GDL and HF/DF lasers.

In case of COIL, the approach consists of employing a tunable diode laser having a peak wavelength around 1310 nm with a bandwidth of 20 nm. Since the iodine atoms also have peak absorption at 1315 nm, the diode laser output is locked to 1315 nm (emission wavelength of $I(^2P_{1/2}) \rightarrow I(^2P_{3/2})$ by passing the probe beam through a heated iodine cell (about 800–850 °C) and observing the peak absorption. During gain measurements, the laser is operated without resonator mirrors and the heated iodine cell is removed from the path of the probe beam, which is passed through the generated active medium in the laser cavity. The chord of observation in each experiment is set by a beam of

He–Ne laser, which is aligned using dichroic mirror (M2) as shown in Fig. 3.8.

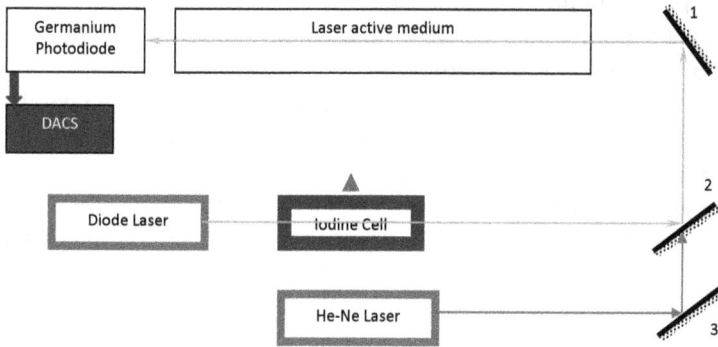

Fig. 3.8. Schematic for gain measurement.

The emission signal along with the signal of the modulated probe beam is observed using a high-gain germanium photodiode used in conjunction with a pre-amplifier and amplifier modules. Since the diode laser has a high scanning rate (100 kHz), the photodiode signals are acquired in a data acquisition system with extremely high sampling rates. The superposed peaks observed over the triangular waveform are due to the stimulated emission of the excited iodine species on encountering photons of a wavelength identical to their characteristic emission wavelength inside the laser cavity. A typical gain curve [21] is shown in Fig. 3.9.

The gain can thus be directly calculated by employing the following relation [20]:

$$I_v = I_o \exp(gL),\qquad(3.26)$$

where g is the small signal gain, I_v is the line center intensity after amplification, I_o is the line center probe laser intensity entering the gain medium and L is the gain length. Further, dividing the determined gain by stimulated emission cross-section would enable determination of excited iodine atom density inside cavity.

There exists a possibility of non-uniform gain over the cross-section of the cavity. Thus in experiment, the cavity is scanned at various points using a motorized platform which was first developed by PSI Inc. and is famously known as Iodine scan diagnostics (ISD) [22] and gain is calculated at these points. For shorter run durations (~ 3 sec) in high power COIL and large cavity dimensions it may be difficult to monitor the gain by scanning method. So array of detectors along with several laser probe beams may be used at several points to estimate the gain simultaneously at several points. It will give a profile of gain in laser cavity. The other technique to measure gain is to divide the cavity into grids. This technique is more costly than the scanning technique as it involves more number of optical components/lasers and detectors. However, this technique is relatively fast and may be easily employed for estimating gain profile in gas lasers. Hence, several variations of the above-discussed setup may be used for determining small signal gain nevertheless the governing principles remain the same.

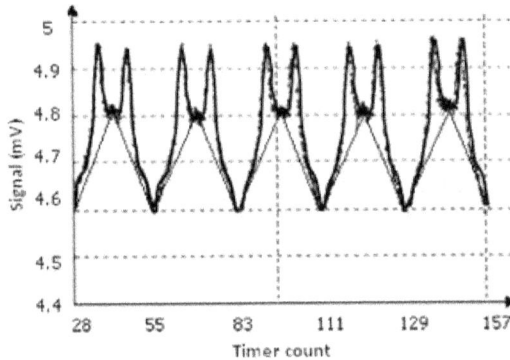

Fig. 3.9. Observed temporal probe beam signal.

In case of HF/DF lasers, a probe beam at the corresponding wavelength is generated by using a suitable diode laser or generating a mini HF/DF laser by dc excitation of subsonic SF_6 followed by injection H or D for generating $HF^*(DF^*)$ [23]. An *InAs PV* detector at 300 K (Judson 512 LD) may be used for detection of amplification of probe beam.

Similar measurement methodology may be employed for gain measurement in case of GDL at 10.6 μm using a small CO_2 laser as the probe beam-generating source.

3.2.1.2 Voigt Profile Method

Voigt Profile, which is a combination of the Lorentz and Doppler profile, has been used for estimation of small signal gain in case of

COIL [24]. COIL as stated earlier operates on the strongest $I(^2P_{1/2},F=3) \rightarrow I(^2P_{3/2},F=4)$ hyperfine component of the spin orbit transition of atomic iodine at the wavelength $\lambda = 1.315$ μm. The broadening of the gain line in the uniform laminar gas stream is due to pressure broadening and Doppler broadening that results in Voigt gain line shape. The small signal gain (SSG) on this transition is equal to:

$$g = \frac{7}{12}\left[N_{I^*} - \frac{1}{2}N_I\right]\frac{A\lambda^2}{8\pi}\left(\frac{2}{\pi\Delta\upsilon}\right), \qquad (3.27)$$

where N_{I^*} is the concentration of excited iodine, N_I is the ground state iodine, λ is the emission wavelength and A is the Einstein stimulated emission coefficient, Δv is the line width corresponding to half maximum between the half power points. However, in case of COIL flows Doppler broadening is significant, hence, Lorentz line function $(2/\pi\Delta v)$ is replaced by Voigt function $\varphi(v)$ [Eq. (3.27), (3.28), (3.29)] for better theoretical accuracy.

$$g(v) = \frac{7}{12}\left[N_{I^*} - \frac{1}{2}N_I\right]\frac{A\lambda^2}{8\pi}\varphi(v) \qquad (3.28)$$

$$\varphi(v) = \sqrt{\frac{\ln(2)}{\pi}}\frac{\Delta v_L}{\pi\Delta v_D}\int_{-\infty}^{\infty}\frac{\exp(-Z^2 4\ln 2/(\Delta v_D)^2)}{(v-v_o-Z)^2 + \left(\frac{\Delta v_L}{2}\right)^2}dZ , \qquad (3.29)$$

where

$Z = \dfrac{u}{\lambda_o}$, u is the absolute gas velocity and λ_o is the centre wavelength

of the probe beam (for COIL it is 1.315 μm).

$\Delta v_D = \dfrac{1}{\lambda_o}\sqrt{\dfrac{8k\ln 2}{M}}\sqrt{T}$ is the Doppler broadening width (FWHM), with

k as Boltzmann constant; T is the static medium temperature in Kelvin, M is the mass of the gas species.

$\Delta v_L = P\left(\dfrac{300}{T}\right)^{\gamma}\Sigma b_i x_i$ is the Pressure broadening width (FWHM), with

b_i is the pressure broadening coefficient at room temperature, x_i is the mole fraction of ith specie, $\gamma = 0.87 \pm 0.13$ [25] for the temperature interval 220-340 K.,

$\Delta N = \left[N_{I^*} - \dfrac{1}{2} N_I \right]$ is the Population inversion. The inverted population is determined by the total concentration of iodine atoms (NI), singlet oxygen yield and static temperature (T) of the medium.

$$\Delta N = \frac{(Y - Y_{th})(K_{eq} + 0.5)}{(K_{eq} - 1)Y + 1} N_I \qquad (3.30)$$

$K_{eq} = 0.75 \; exp(402/T)$ is the equilibrium constant for pumping of iodine atom by singlet oxygen molecule and $Y_{th} = (2K_{eq}+1)^{-1}$ is the threshold singlet oxygen yield at which there is no population inversion.

Thus, small signal gain as function of scanned frequency may be determined using the above stated relations. The simplest case is the one in which the probe beam enters the medium normal to the flow, the incident and reflected beam are slightly shifted to avoid overlap. During mathematical deconvolution gain as function of scanned frequency is considered to be given by single Voigt function and the overall gain, g (X) is taken as an average of $g_1(X)$ and $g_2(X)$. Since the probe laser has a Lorenzian line shape with Δv_p, the observed spectrum is a convolution of the probe and the gain spectrum. Therefore, the observed width of the Lorenzian line shape is $\Delta v_D + \Delta v_p$. Further, mathematical processing of these functions allows the determination of the peak SSG g_1 (0), g_2 (0) as well as Δv, Δv_D, Δv_L. An example of the gain profile is shown in Fig. 3.10.

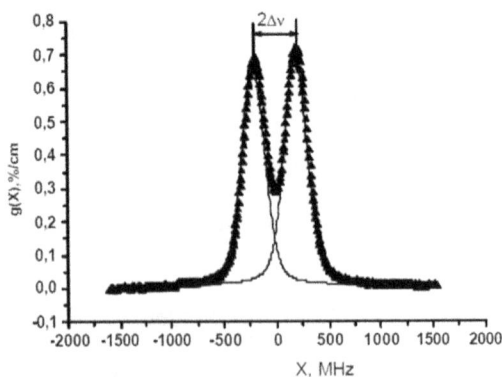

Fig. 3.10. SSG spectrum for probe laser beam, triangles represents the experimental values and solid line is Voigt function fit on g(X).

3.2.2 Mach Number Diagnostics

Gas chemical lasers efficiency depends not only on the effective transportation of the pumping source and lasing species into the laser resonator for proper mixing but also on the flow conditions of these species. The lasing conditions inside the cavity or resonator are achieved employing supersonic nozzles, the performance of which is governed by the upstream gas flow parameters. The extent of population inversion i.e. small signal gain inside the laser resonator is a function of temperature and so depends to a large extent on the Mach number in the cavity.

In the development of a gas laser system, Mach number needs to be monitored continuously during the run for performance optimization. The high-resolution determination of Mach number during run is thus a critical diagnostics representing the lasing medium behavior in the cavity. Further, the performance analysis of the system requires their data storage in graphical form and online utilization of these parameters for evaluation of various other critical parameters.

In case of CO_2 GDL, typical laser cavity pressure is 35-40 torr and Mach number is of the order of 4-5 whereas laser cavity pressure of HF/DF laser is 4-6 torr typically and Mach number of lasing flow is ~3-5. However COIL can operate in subsonic as well as supersonic regime. The typical laser cavity pressure is 3 torr and Mach number of lasing flow may be 0.8-0.9 (for subsonic COIL) and 1.5-2 (for supersonic COIL).

Thus Mach number determination unit must be able to estimate Mach number ranging from subsonic region to supersonic region. The first among the various methods that may be used is Pitot static tube method.

3.2.2.1 Pitot Static Tube Method

The Mach number can be estimated on-line as a function of time, from the measurements of static and stagnation pressure values at the cavity. The operation schematic for Pitot static tube is shown in Fig. 3.11.

As is evident that in subsonic flow, the flow velocity reduces to zero at the tip of the tube and the surrounding fluid is at static pressure both of which are measured simultaneously.

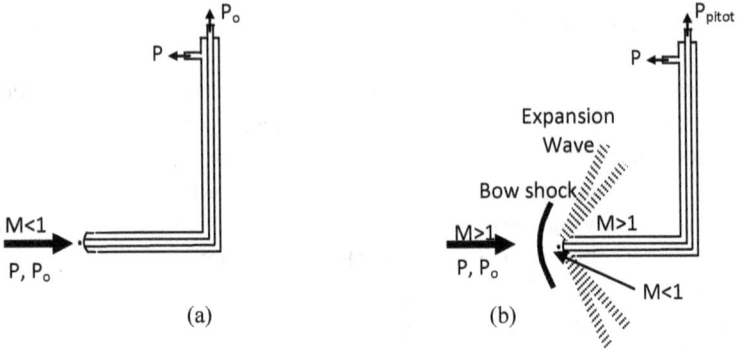

Fig. 3.11. Operation schematic for Pitot static tube
(a) Subsonic case; (b) Supersonic case.

Thus, for the case of subsonic and sonic flow i.e. $M \leq 1$, the static pressure (P), Pitot pressure ($P_{Pitot} = P_o$) and flow Mach number (M) are related for a near isentropic flow i.e. for Pitot tube like measurements as:

$$P_o = P \left[1 + \frac{\gamma - 1}{2} M^2 \right]^{\frac{\gamma}{\gamma - 1}}, \qquad (3.31)$$

whereas, for a supersonic flow i.e. $M > 1$, the Pitot tube pressure corresponds to a total pressure across a curved or bow shock which is considered equivalent to a mini normal shock as the Pitot tube size is quite small. Also, expansion waves emanate from the edge of the tube thus allowing the flow to again expand to original Mach number around the tube, which is measured by the static pressure measurement holes on the side of the tube. Since there is a stagnation pressure loss across this shock resulting in $P_{Pitot} < P_o$ which needs to be accounted for in the Mach number determinations, thus, the Rayleigh -Pitot relation [26], true for a normal shock, is used for the Mach number estimation by iterative method from the measurements of Pitot (P_{pitot}) and static (P_{cavity}) pressures,

$$\frac{P_{Pitot}}{P_{cavity}} = \frac{\left[\frac{\gamma + 1}{2} M^2 \right]^{\frac{\gamma}{\gamma - 1}}}{\left[\frac{2\gamma}{\gamma + 1} M^2 - \frac{\gamma - 1}{\gamma + 1} \right]^{\frac{1}{\gamma - 1}}} \qquad (3.32)$$

Thus, it needs a data loop, which is based on general relations for gas dynamic flow true for both subsonic and supersonic regimes. The computation loop for the Mach number would typically involve the sequence as shown in Fig. 3.12. Pitot static tube method is one of the important methods, which is used commonly on account of its simplicity and its applicability to all kinds of supersonic flows. However, one of the drawbacks is that it is intrusive and may influence the flow conditions. Hence, other methods mostly optical based like Laser Doppler Velocimetry (LDV) or Voigt profile may also be used since they are non intrusive in nature.

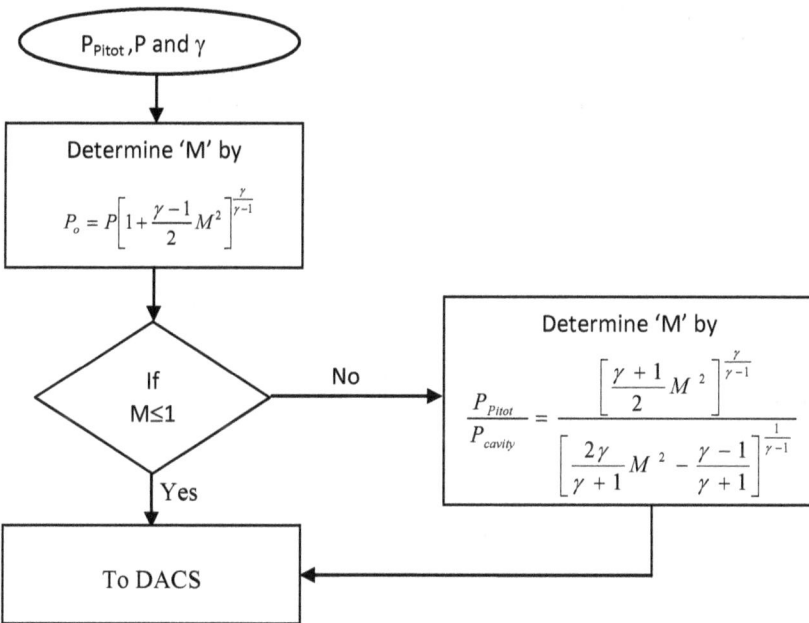

Fig. 3.12. Computation loop for Mach Number Estimation.

3.2.2.2 Laser Doppler Velocimetry (LDV)

Laser Doppler Velocimetry (LDV) is a technique used to measure the instantaneous velocity of a flow field. This technique, like PIV is non-intrusive and can measure all the three velocity components. The laser Doppler velocimeter sends a monochromatic laser beam toward the target and collects the reflected radiation. According to the Doppler effect, the change in wavelength of the reflected radiation is a function of the targeted object's relative velocity. Thus, the velocity of the object can be obtained by measuring the change in wavelength of the reflected laser light, which is done by forming an interference fringe pattern (i.e. superimpose the original and reflected signals). This is the basis for LDV. The schematic is shown in Fig. 3.13.

A flow is seeded with small, neutrally buoyant tracer particles that scatter light. The particles are illuminated by a known frequency of laser light. The scattered light is detected by a photomultiplier tube (PMT), an instrument that generates a current in proportion to absorbed photon energy, and then amplifies that current. The difference between the incident and scattered light frequencies is called the Doppler shift. By analyzing the Doppler-equivalent frequency of the laser light scattered (intensity modulations within the crossed-beam probe volume) by the seeded particles within the flow, the local velocity of the fluid can be determined.

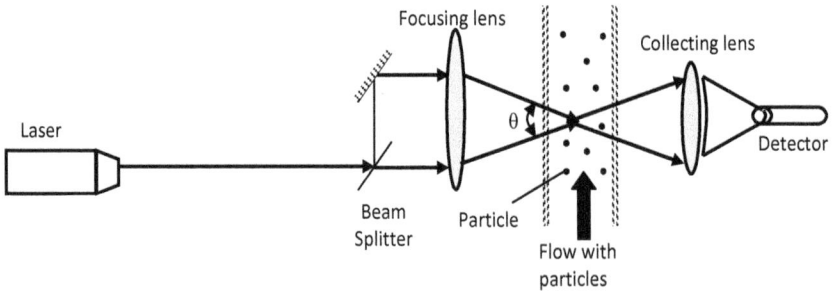

Fig. 3.13. Schematic for Laser Doppler Velocimetery.

It basically uses the principle of optical beating, and the particle velocity may be calculated by using the understated expression,

$$f_d = \frac{2 \sin\left(\dfrac{\theta}{2}\right)}{\lambda} u, \qquad (3.33)$$

where f_d is the Doppler shift between the incident and the scattered frequencies, λ is the wavelength of the probe beam which is typically a He-Ne laser at 632.8 nm, θ is the angle between the beams forming the probe volume and u is the fluid particle velocity.

In case of gas lasers there is a predominantly single direction of fluid velocity but multiple beam LDV is capable of measuring all three component of velocities as well. LDV enables velocity measurement, but to be able to determine Mach number, one needs to know the cavity temperature which may be correlated from temperature measurements carried out at generator exit or by performing specific temperature measurements inside cavity employing intrusive or non–intrusive techniques.

This type of a measurement for flow velocity is possible in case of GDL, which is primarily combustion based and has sufficient particulate matter for producing Doppler scattered light. Clear gas flows without presence of any particulate matter as encountered in COIL and HF/DF flows will require introduction of tracer particles for using LDV, which is generally not a preferred methodology.

3.2.2.3 Voigt Profile Method

The details of Voigt profile fitting have been discussed in the session on small signal gain. The measured central frequency of small signal gain for the beam directed at an angle θ with respect to the normal to the flow velocity (shown in Fig. 3.14) is shifted by $\Delta \upsilon$, which is given by [24],

$$\Delta v = \frac{2 \sin(\theta)}{\lambda} u,\qquad (3.34)$$

where u is the absolute gas velocity. The gas temperature may be easily determined from the Doppler broadening expression given earlier under SSG. Correspondingly, Mach number may be calculated using the following expression,

$$M = u \sqrt{\frac{\mu}{kRT}},\qquad (3.35)$$

where μ is the molecular weight, $k = 1.4$ typically in case of chemical gas lasers, R is the universal gas constant.

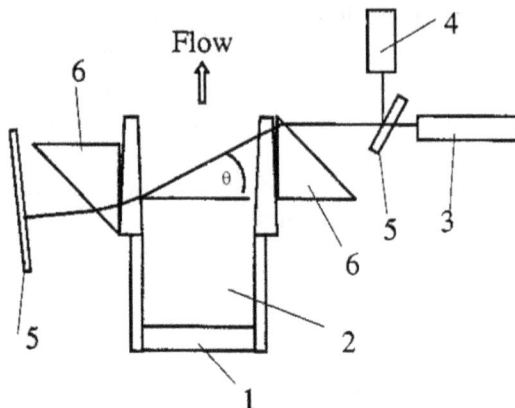

Fig. 3.14. Schematic of setup; 1- Nozzle bank; 2- Mixing chamber; 3- Probe laser; 4 - Photo detector; 5- Mirrors; 6-Optical prisms.

3.2.3 Medium Homogeneity

An important factor affecting the gain in gas lasers is the homogeneity of the medium. In solid-state lasers, with the advancement of fabrication technology it is now possible to dope lasing elements (e.g. Nd in YAG matrix) in the host material fairly uniformly. Thus, the gain of solid-state lasers remains constant throughout the material cross-section. However, this is not the case with gas lasers. In these lasers several gases flow in the cavity e.g. CO_2, N_2, air in CO_2 GDL system, I_2, Singlet oxygen, Cl_2, N_2 and water vapors in COIL, HF/DF, fluorine, N_2, H_2 in HF/DF laser. The proper mixing of gases is critical for achieving optimum power. In order to estimate homogeneity in laser chamber, it is required to implement suitable measurement technique. Let us take COIL case further in detail to explain the need for homogeneity estimation. In COIL, Singlet oxygen is mixed with Iodine molecules in a laser cavity with the help of supersonic nozzle. In addition, nitrogen is also supplied as buffer gas with singlet oxygen not only to establish pressure conditions as well as to dilute singlet oxygen for preventing quenching losses. Since Iodine molecule is heavier, it is also transported with the help of buffer nitrogen. The singlet oxygen serves two purposes: first it dissociates iodine molecule into iodine atom and then it further excites iodine atom to excited state and produces population inversion for lasing action. Hence, efficient mixing of singlet oxygen with Iodine is thus of prime importance for obtaining maximum gain.

Mixing is one parameter, which is apparently difficult to quantify but with its increasing importance in various fields, effective techniques have been developed for determining the extent of mixing between multiple gas streams. The subsequent section deals with the possible techniques for estimation of mixing in case of chemical gas lasers.

3.2.3.1 Gain Mapping

Measurement of small signal gain using iodine scans diagnostics (ISD) or Voigt profile method as explained earlier enables mapping the SSG data across the aperture of the resonator window in the direction transverse the flow. The obtained data provides qualitative information regarding mixing or medium homogeneity as the gain is high in regions where the mixing is good and poor in regions of insufficient mixing.

The quantification of mixing efficiency on the basis of SSG mapping transverse to the flow direction has been suggested by Rosenwaks *et. al.* [27] in case of COIL, using the relation stated below:

$$\eta_{mix} = \frac{\int_0^H g(y)\,dy}{g_{max}\,H}, \qquad (3.36)$$

where g (y) is the gain at certain y location, g_{max} is the maximum observed gain, and H is the nozzle height. The typical variation of gain for a slit nozzle, advanced and winglet nozzle case is shown in Fig. 3.15.

It is clear that qualitatively the gain profile in the transverse direction is much broader for winglet nozzle than in the case of conventional slit nozzle or in advanced nozzle case. The mixing efficiencies are nearly 49 %, 60 % and 74 % [21] for the slit, advanced and winglet nozzles respectively.

The above methodology may be easily extended for GDL and HF/DF lasers as well by generating a transverse SSG map at various locations along the flow direction. This would also enable one to ascertain the location from the nozzle throat where the medium homogeneity is the best and mixing is complete.

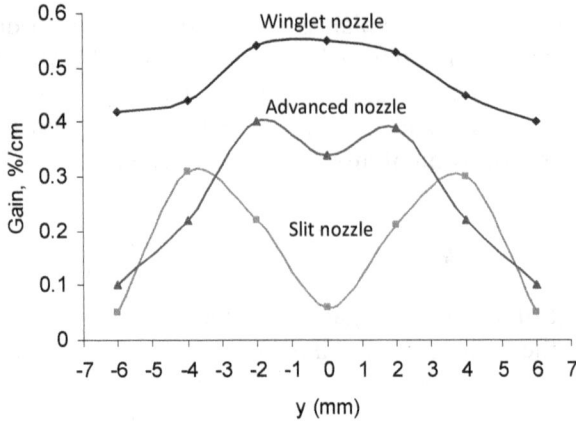

Fig. 3.15. Gain profile transverse to both the direction of flow
and to the optic axis at the location of the optical plane.

3.2.3.2 Optical Interferometer

Optical interferometer is one amongst the direct measurement techniques now available to measure the homogeneity of the medium. It is one of the most accurate techniques as it is based on interferometer principle. We know that light travels with different velocity in different mediums and hence travels different distance in a given time. This produces a phase difference in the beams travelling in two different mediums. When these beams travelling through different medium are allowed to meet, they produce interference and fringe patterns are obtained. The fringe width is directly proportional to the phase shift or the refractive index of the medium. A typical scheme for measurement of homogeneity of lasing medium in chemical gas laser is based on Mach- Zehnder interferometer, which is shown in Fig. 3.16. It primarily maps the variation in density of the cavity medium depicting how well various species are dispersed in the medium. A near uniform map of density represents good medium homogeneity and efficient mixing of species.

In GDL, this principle may be applied to investigate the homogeneity of the medium. In experiments, a laser probe beam is divided into two paths with the help of beam splitter. First beam is allowed to pass through air and the second beam scans the cavity. These two beams meet at detector and produce interference. Since the medium is gas, the refractive index strongly depends on the density of medium. Thus, the fringe width at a point gives direct information about the density of medium. A scan can thus produce the entire density profile of the medium and regions of low density (refractive index) performance can be identified and rectified by changing the nozzle configuration or upstream flow conditions.

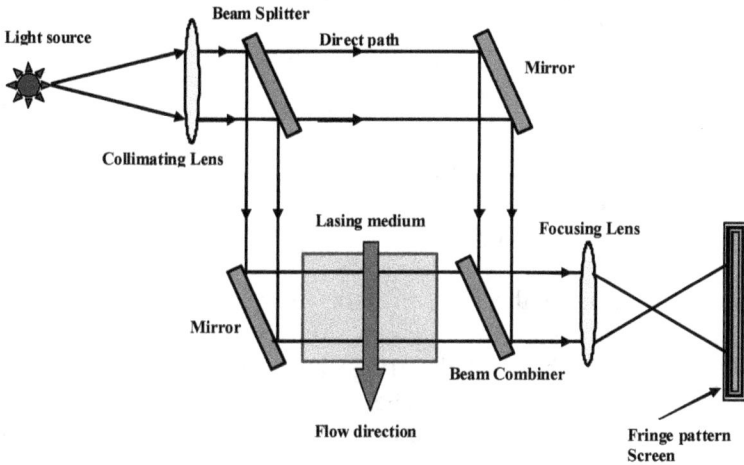

Fig. 3.16. Optical interferometer based homogeneity measurement.

The interference pattern (Interferogram) consists of bright and dark fringes. If the light waves passing through the test section are on dark bands, they are out of phase with the waves in open arm (in air) by 1/2, 3/2, 5/2.....of the wavelength (λ). In this way the separation between two adjacent fringes on dark band will be just one wavelength long. Thus the time difference for light beam to pass through two different locations in laser medium 'a' and 'b' forming adjacent dark fringe patterns will be,

$$t_b - t_a = \frac{L}{c_b} - \frac{L}{c_b} = \frac{\lambda_{vac}}{c_{vac}},$$ (3.37)

where c_a and c_b are the corresponding velocities and L is the length of the test section along the light direction.

Now, we know that the refractive index of the medium is the ratio of velocity of light in different mediums, and using the empirical formula which relates refractive index (n) with density (ρ) of the medium as

$$n_b - n_a = \frac{c_{vac}}{L}\left(\frac{L}{c_a} - \frac{L}{c_b}\right) = \rho_b - \rho_a = \frac{\lambda_{vac}}{LK}, \qquad (3.38)$$

where K is the Gladstone-Dale constant for a particular gas. Further, if the density of medium were to change uniformly from condition 1 to condition 2, the fringe shift [28] is related to density as

$$\rho_1 - \rho_2 = \frac{\lambda_{vac}}{LK}\frac{l}{d}, \qquad (3.39)$$

where l is the distance shifted by a dark fringe in passing from condition '1' to condition '2' and 'd' is the distance between dark fringes in reference condition.

Thus, by measuring the shift in dark fringes, one can calculate the density distribution of the medium. Thus, the uniform spacing and parallelism of the band of the fringe pattern is representative of the homogeneity of the medium. The density of the medium may then easily be mapped using various available image processing software.

As far as, low-pressure gas lasers such as COIL, is concerned, implementation of this technique is relatively challenging. This is mainly because the accuracy of the technique is limited by dn/dρ i.e. the change in refractive index with respect to the density of the medium. However, in case of CO_2 GDL, density of medium is ~10 times more than in case of COIL, hence this technique may be more suitably applied in CO_2 GDL for homogeneity measurement of gain medium.

3.2.3.3 Laser Induced Fluorescence (LIF) / Planar-LIF (PLIF)

Laser Induced Fluorescence (LIF) and now -a- days Planar laser-induced fluorescence (planar-LIF) are both common optical measuring techniques used to measure instant whole-field concentration or temperature maps in various flow fields including supersonic flows. These are being extensively used to study complex fluid dynamic domains such as premixed combustion, turbulent mixing etc.

Chemical gas lasers as has been stated earlier involve mixing of various fluid streams. Thus, LIF may be used to determine the uniformity of the lasing species in direction of the optical axis. PLIF goes a step further in providing a map of the manner of distribution of the specie being detected over a complete plane in the flow field. PLIF visualization of say the injected flow can enable identification of turbulent structures, penetration distance of the injected (secondary) flow into the primary flow, and relative concentration of the secondary to primary flows.

Considering, an example of LIF/PLIF in case of COIL [29, 30], which has an advantage that the basic lasing species i.e. Iodine molecules itself serve as a fluorescence marker. As reported in literature researchers have used two options for pumping iodine molecules from X $^1\Sigma$ state to B $^3\Pi$ state, with the excited iodine molecules emitting yellow light. One of the methods is to use Ar-laser [29] emitting at 514.5 nm and the other is to use Nd:YAG laser emitting at 532 nm to pump a tunable dye laser with Rohdamine 6G emitting over a range of 559-576 nm [31, 32]. The excited iodine molecules show peak fluorescence at 565 nm. According, to Sakurai [33] the fluorescence of iodine at pressures those experienced in the cavity is sustained for 1.5 μs, which should be greater than the typical gating time for cameras (1 μs) allowing the cameras to take images when laser pulse is off but the fluorescence is on.

In case of LIF, the beam is passed in direction perpendicular to the flow and LIF picture is recorded by a Kodak film or some other similar manner in dark conditions with an optical filter to remove scattering radiation from the incident beam. The schematic of the same is shown in Fig. 3.17.

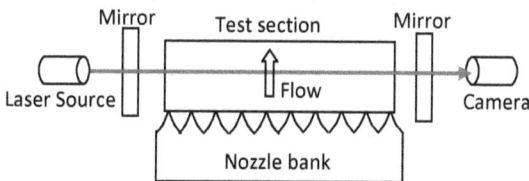

Fig. 3.17. Laser Induced Fluorescence schematic.

In case of PLIF, the beam is passed through a set of spherical and cylindrical lenses to form a laser sheet in a direction perpendicular to the flow. The PLIF images are then recorded by using an amplified CCD camera preferably perpendicular to the sheet. Incase of oblique imaging, applicable perspective correction may be carried out. The schematic of the same is shown in Fig. 3.18.

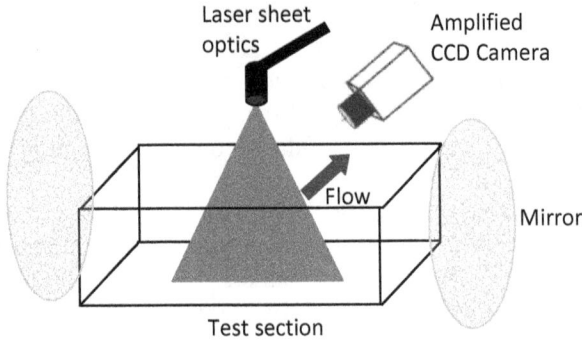

Fig. 3.18. Planar laser induced fluorescence (PLIF) schematic.

The fluorescence signal (S_f) is given by the expression stated below

$$S_f = \frac{E\lambda}{hc}\frac{\chi p}{kT}L\sigma(\lambda,T)\phi(\lambda,p,T)\frac{\Omega}{4\pi}\eta, \qquad (3.40)$$

where E is the incident laser energy, λ is the laser wavelength, p is the total pressure, T is the temperature, χ is the mole fraction of fluorescing tracer, L is the length of the illuminated volume, Ω is the detector collection angle, and η is the detector collection efficiency. The fluorescence quantum yield (FQY) ϕ and absorption cross-section σ are photo physical parameters specific to each tracer.

In order to quantify the mixing behavior of the secondary and primary flows, histograms of the fluorescence intensity in the stack of statistically independent ensemble images are required to be created using image processing software such as DaVIS. Histograms display the probability density of the intensity. In each image, the intensity is normalized by the maximum intensity. Initially, a high intensity secondary flow structure may be seen close to the point of injection. As one moves downstream an intermediate intensity well mixed flow is generally observed. Also, the flow becomes fully developed at certain downstream location, after which the mixing quality does not increase drastically with downstream distance.

In case of HF/DF and GDL lasers LIF/PLIF studies may be performed for determining medium homogeneity by introducing suitable fluorescent markers.

3.3 Laser Output Power and Pulse Shape Measurement

The aim of development of gas lasers is to obtain a high laser power (ranging from few hundred watts to several hundred kW or even MW). It is important to know the output power and its pulse shape for the characterization of gas lasers. The present session discusses the two aspects in conjunction as these two are closely related to one another,

In order to fully characterize a laser one must measure the energy and the power or, more specially, the radiant energy in joules (J) and radiant power or energy flux in $J\ s^{-1}$. Due to the nature and diversity of lasers which operate from the deep ultraviolet to the infrared, from continuous output to picoseconds pulses and from microwatts to megawatts, energy measuring devices encompass a wide range of possibilities [34]. Detectors are sensitive to the spectrum of the beam and speed of energy deposition in their conversion of energy to a useful readable or recordable quantity that can be interpreted as energy or power. The reliable detection of the optical radiation is an essential element of laser technology.

Laser energy measurement can be based on photoelectric, photochemical or calorimeter method. Photodetector may be employed either to sample laser output itself or to detect radiation produced by some other system as a result of laser excitation [35]. By far the most common method by which intensity of laser light is measured is by conversion to a real–time electrical signal. Detection of optical radiation relies on either detecting the light due to absorption of its energy by a body and observing the change in its temperature (photothermal), or observing the effect on the detector when individual photons are detected (photoelectric). Generally, semiconductors are used in photoemission and photothermal detection mode.

Thermal detectors contain two principle elements: an absorption region, the temperature of which is a function of the incident optical power, and a transducer to convert the temperature variation in the detector to an electrical signal. The most commonly used types of transducers are thermocouples and thermopiles; bolometers and thermistors; and pyroelectric detectors. Thermal detectors are most commonly used for detection of optical radiation at wavelengths in the mid infrared and longer, as the thermal detection process is slow and inefficient compared with the photoelectric processes, and the latter are usually preferred within their range of sensitivity. However, the ability of thermal detectors to provide very uniform detection over a wide range of optical wavelengths has proved useful also for radiometric calibration purposes.

Both photoelectric and thermal detectors cannot be used for measurement of high laser power because they cannot handle very high power. One way is to use beam splitter and measure a fraction of power and estimate the total power but it is not quite an accurate method. The other way is to use calorimetric principle [36] for laser power measurement. The laser power is allowed to fall on a conical dump made of black anodized copper material for maximum absorption with minimum loss. A series of K type thermocouples are fixed at the other surface of the cone for measuring the temperature rise. Since the temperature rise is proportional to the energy fall on the conical dump, the total energy (Q) fall on the power meter is estimated using the relation,

$$Q = mc\Delta T \,, \tag{3.41}$$

where m is the mass of the calorimeter, 'c' is the specific heat capacity of the calorimeter material and 'ΔT' is the temperature rise. The calorimeter is calibrated for determining the correct heat capacity factor or the calibration factor 'mc' and the total energy dumped in to the calorimeter during the run is directly estimated using this calibration factor and the measured temperature rise.

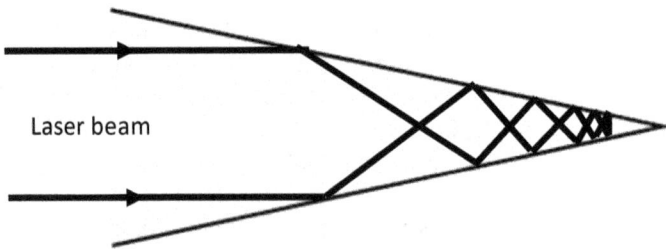

Fig. 3.19. Radiation paths in a cone calorimeter.

It may be seen from Fig. 3.19 that the reflection paths lead to concentration of the beam energy toward the tip of the cone. Thus, in order to have uniform temperature rise across the entire body of the calorimeter the thermal conductivity of the material should be high. Fig. 3.20 shows the photograph of a typical cone calorimeter.

Fig. 3.20. Hardware of cone calorimeter used for COIL application.

The major problem of laser beam calorimetry is that of ensuring conversion of all of the incident radiation to heat. Here, the problem differs for cw lasers and for pulsed lasers. In general, the peak powers (per unit area) of cw lasers are low enough for a variety of blackening materials to be used to obtain nearly complete absorption at a surface. But pulsed lasers, particularly the Q-switched and mode-locked types, may have peak powers several orders of magnitude greater than those of cw lasers. These high powers may damage a blackened surface so that the absorptivity rapidly decreases; they may lead to transient surface temperatures so high that energy is lost by volatilization of material from the surface of the calorimeter, or by thermal radiation from surfaces of low thermal conductivity.

The problem may partially be solved by using a surface of lower absorption arranged in a geometrical configuration (cone or hollow sphere), so that incident radiation must undergo many reflections before escaping from the calorimeter. The surface may then be a bright metal, which is much less susceptible to damage than black paint; and because of the multiple reflection path, the overall reflectivity from the calorimeter is insensitive to changes in the single-surface reflectivity.

Once the mass of calorimeter is decided, one can always measure the change in temperature due to laser energy falling on it. Thus using relation 3.41, laser energy can be estimated. Laser power can subsequently be calculated by taking a ratio of energy with laser pulse duration.

Apart from direct measuring cone calorimeter, one may employ an indirect measuring cone calorimeter. Such calorimeters typically employ a coolant such as water to keep them at a steady temperature for a given incident heat flux by rejecting the heat to the coolant. Thus, measuring the inlet and outlet temperature of the coolant under stable conditions one can easily estimate the total energy dumped. Finally, from the known duration of the laser pulse power may be estimated.

The laser pulse shape and pulse duration can be observed by collecting the scattered photon from the rear mirror (reflecting mirror) of the laser by introducing a scattering medium. The pulse duration measurement is important for power estimation. In case of CO_2 GDL, laser emits at 10.6 μm, which can be detected using Mercury cadmium telluride detector. HF/DF laser wavelength is 2.7-3.7 μm, detection of which may be carried out by using Indium antimonide detector. COIL (1.315 μm) may be detected with the help of Germanium photodiode. The FWHM of the temporal profile of the laser pulse is used for the power output estimation using the net temperature rise observed in the calorimeter.

3.4 Selection of Diagnostic Techniques

A brief discussion on the most applicable techniques for measuring key parameters in different scenarios for chemical gas lasers is undertaken in the present session.

Optical emission and absorption methods have been used extensively for non intrusive measurement of concentration of various species. Typically, absorption based methods are apparently more preferred mode of measurements and provide greater accuracy. With the advent of diode laser based methods carrying out these measurements has become easier for species which were measured using optical emission by pumping specific transitions difficult to achieve using broad band sources. These methods also lead to less error in the measurement of species.

Thus, if we consider a case of COIL laser of reasonable power, chlorine utilization, and water vapor and ground state oxygen concentration is measured using diode laser based absorption methods. Singlet oxygen yield may then be estimated as the flow residual by using partial pressure relations. Similarly iodine concentration measurements may again be made using normal optical absorption. However, if we consider a μCOIL system, measurements of ground state oxygen and iodine concentration would probably require CRDS method.

If we were to examine the case of small signal gain, a scan diagnostics based on Voigt profile method would be the best. This is because it would enable determination of peak gain (with frequency) at several spatial locations. Further, it may be used to estimate cavity temperature and corresponding Mach number of the flow. Also, by using spatial gain mapping technique efficiency of mixing of the multiple gas streams may also be determined. The technique does provide precise estimate for gain and Mach number but as far as quantification of mixing is concerned it is prone to relatively higher errors as compared to a PLIF method. PLIF is the best technique available as far as study of mixing is concerned and is finding increased usage in studies of chemical laser flow fields.

If one is to take in to account costs as well, then in case of Mach number the most economical method of is the Pitot –static tube method and with miniaturization of pressure probes the error due to flow intrusion have also diminished. Thus, ultimate choices depend on the available resources and the desired level measurement accuracy.

In case of GDL due to presence of trace particles and higher pressure of the medium Mach number may be measured using LDV and medium homogeneity is quantified using optical interferometer.

Measurement of power is done using calorimetric method as both photoelectric and thermal detectors cannot be used at high power levels. Direct methods may be used for not very high power levels of the order of 20 kW. In case of very high power levels, upwards of 50 kW indirect calorimeters using water as a coolant are generally employed.

References

[1] S. J. Davis, Historical Perspective of COIL Diagnostics, *SPIE*, 4631, 2002, pp. 60.
[2] D. J. Benard, W. E. McDermott, N. R. Pchelkin, R. R. Bousek, Efficient Operation of a 100W Transverse Flow Oxygen Iodine Chemical Laser, *Applied Physics Letters*, 34, 1, 1979, p 40.
[3] Marcelle V. Zagidullin, Valery D. Nikolaev, Michael I. Svistun, Vladimir S. Safonov and N. I. Ufimtsev, The Study of Buffer Gas Mixing with

Active Gas on Chemical Oxy-Iodine Laser Performance with Jet SOG, *SPIE*, 2702, 1996, p. 310.

[4] R. Rajesh, Gaurav Singhal, Mainuddin, R. K. Tyagi, A. L. Dawar, High throughput jet singlet oxygen generator for multi kilowatt SCOIL, *Journal of Optics and Laser Technology*, Vol. 42, 2010, p 580.

[5] O. Spalek and Jarmila Kodymova, Singlet Oxygen Generator for Supersonic Chemical Oxygen Iodine Laser - Parametric Study and Recovery of Chemicals, *SPIE*, 2987, 1997, p. 131.

[6] R. K. Tyagi, R. Rajesh, Gaurav Singhal, A. L. Dawar, M. Endo, Design and realization of a 500 W class jet type singlet oxygen generator with angular exit, *SPIE*, 4971, 2003, p. 13.

[7] J. Kodymova and O. Spalek, Performance characteristics of jet type generator of singlet oxygen for supersonic chemical oxygen iodine laser, *Jpn. J. Appl. Phys.*, Vol. 37, 1, 1998, p 117.

[8] R. Rajesh, Gaurav Singhal, Mainuddin, R. K. Tyagi, and A. L. Dawar, High throughput jet singlet oxygen generator for multi kilowatt SCOIL, *Opt. Laser Technol.*, Vol. 42, 4, 2010, p. 580.

[9] C. Pradayrol, A. M. Casanovas, I. Deharo, J. P. Guelfucci and J. Casanovas, Absorption Coefficients of SF6, SF4, SOF2 and S02F2 in the Vacuum Ultraviolet, *J. Phys. III France*, Vol. 6, 1996, p 603.

[10] M. V. Zagidullin, V. D. Nikolaev, M. I. Svistun, N. A. Khvatov, N. I. Ufimtsev, Highly efficient supersonic chemical oxygen iodine laser with a chlorine flow rate of 10 mmol/s, *Quantum Electronics*, Vol. 27, 1997, p. 195.

[11] R. K. Tyagi, R. Rajesh, Gaurav Singhal, Mainuddin, A. L. Dawar and M. Endo, Parametric studies of Supersonic COIL with Angular Jet Singlet Oxygen Generator, *Infrared Science and Technology*, Vol. 44, 2003, p. 271.

[12] P. Sulzer and K. Wieland, Calculation of the intensity distribution of a continuous Iodine absorption spectrum and its dependence on wave number and temperature, *Helv. Phys. Acta*, Vol. 25, 8, 1952, p. 653.

[13] Mainuddin, M. T. Beg, Moinuddin, R. K. Tyagi, R. Rajesh, Gaurav Singhal and A. L. Dawar, Optical spectroscopic based In-line iodine flow measurement system-an application to COIL, *Sensors and Actuators*: B, Vol. 109, 2005, p. 375.

[14] Mainuddin, Gaurav Singhal, R. K. Tyagi, and A. K. Maini, Diagnostics and data acquisition for chemical oxygen iodine laser, *IEEE Transactions on Instrumentation and Measurement*, 61, 6, 2012, p. 1747.

[15] M. G. Allen, K. L. Carleton, S. J. Davis, W. J. Kessler, and K. R. McManus, Diode laser- based measurements of water vapor and ground state oxygen in chemical oxygen iodine lasers, *AIAA Paper*, 94-2433, 1994.

[16] W. Zhao, F. Sang, L. Duo, F. Chen, Y. Zhang and B. Fang, Measurement of chemical oxygen –iodine laser singlet oxygen generator parameter using Raman Spectroscopy, *SPIE*, 4631, 2002, p. 161.

[17] M. D. Wheeler, S. M. Newman, A. J. Orr-Ewing and M. N. R. Ashfold, Cavity ring-down spectroscopy, *J. Chem. Soc., Faraday Trans.*, 94, 3, 1998, p. 337.

[18] G. Berden, R. Peeters, and G. Meijer, Cavity ring-Down spectroscopy: Experimental schemes and applications, *Int. Rev. Phys. Chem.*, 19, 4, 2000, p. 565.

[19] K. B. Hewett, J. E. McCord, Manish Gupta, and Thomas Owano, Measuring the yield of singlet oxygen in a chemical oxygen iodine laser, *SPIE*, 6346, 2007, p. 63460E.

[20] Jarmila, Kodymova and Otomar Spalek, A contribution of COIL Laboratory in Prague to the Chemical Oxygen-Iodine Laser research and development, *SPIE*, 4631, 2002, p. 86.

[21] Gaurav Singhal, P. M. V. Subbarao, A. L. Dawar, M. Endo, Computation of streamwise vorticity in a compressible flow of a winglet nozzle based COIL device, *Optics and Laser Technology*, Vol. 40, 2008, p. 64.

[22] S. J. Davis, M. G. Allen, W. J. Kessler, and K. R. McManus, M. F. Miller and P. A. Mulhall, Diode laser- based sensors for chemical oxygen iodine lasers, *SPIE*, 2702, 1996, p. 195.

[23] K. T. Yano, H. M. Bobitch, Small signal gain measurement in a cw Chemical Laser, *IEEE Journal of Quantum Electronics*, Vol. 14, 1, 1978, p. 12.

[24] V. D. Nikolaev, M. V. Zagidullin, M. I. Svistun, B. T. Anderson, R. F. Tate, G. D. Hager, Results of Small-Signal Gain Measurements on a Supersonic Chemical Oxygen Iodine Laser With an Advanced Nozzle Bank, *IEEE Journal of Quantum Electronics*, Vol. 38, 5, 2005, p. 421.

[25] V. D. Nikolaev, M. I. Svistun, M. V. Zagidullin, and N. A. Khvatov, The temperature dependence of pressure broadening of $^2P_{1/2} - ^2P_{3/2}$ transition of atomic iodine, *Quantum Electron,* Vol. 31, 2001, p. 373.

[26] Mainuddin, R. K. Tyagi, R. Rajesh, Gaurav Singhal and A. L. Dawar, Real-time data acquisition and control system for a chemical oxygen-iodine laser, *Measurement Science and Technology*, 14, 2003, p. 1364.

[27] S. Rosenwaks, B. D. Barmashenko, E. Bruins, D. Furman, V. Rybalkin and A. Katz, Gain and Temperature in a slit nozzle supersonic chemical oxygen iodine laser with transonic and supersonic injection of iodine, *SPIE,* 4631, p. 2001.

[28] E. Rathakrishnan, Gas Dynamics, Second Edition, *Prentice Hall of India*, 2008.

[29] V. D. Nikolaev, M. V. Zagidullin, G. D. Hager, T. J. Madden, An efficient supersonic COIL with more than 200 torr of the total pressure in the active medium, *AIAA Paper,* 2427, 2000.

[30] C. A. Noren, C. R. Truman, P. V. Vorobieff, PLIF Flow Visualization of a Supersonic Injection COIL Nozzle, *AIAA Paper*, 5388, 2005.

[31] C. A. Noren, C. R. Truman, P. V. Vorobieff, PLIF Flow Visualization and Quantitative Mixing Measurements of Supersonic Injection Nozzle, *AIAA Paper*, 2895, 2006.

[32] C. A. Noren, C. R. Truman, P. V. Vorobieff, Quantitative Mixing Measurements of Supersonic Injection COIL Nozzle with Trip Jets, *AIAA Paper*, 3881, 2008.

[33] K. Sakurai, G. Capelle, H. P. Broida, Measurements of Lifetimes and Quenching Cross Sections of the B $^3\Pi_{ou}$ state of iodine using tunable dye laser, *Journal of Chemical Physics*, Vol. 54, 3, 1971, p. 1220.

[34] W. Budde, Physical Detectors of Optical Radiation, *Academic*, New York, 1983.

[35] K. R. Nambiar, Lasers: principles, types and applications, *New Age International (P) Limited,* 2004.

[36] Stuart R Gunn, Lawrence Livermore Laboratory, California, USA, Review Article: Calorimetric measurement of laser energy and power, *Journal of Physics E: Scientific Instruments*, Vol. 6, 1973, p. 105.

Chapter 4

Signal Conditioning

Signal conditioning is an essential aspect of any measurement system. Signal conditioning is required for manipulating a signal in such a way that it meets the requirement of front end processing. In chemical gas lasers, next stage is mainly analog to digital conversion from data acquisition point of view. Transducers [1-2] must be designed not only to be interfaced with either display devices or Data acquisition modules or other circuits, but also they must be able to transmit signals over a variety of communication channels (wired or wireless). Transducers must be able to cope with various types of noise and interference during data transmission. The change of energy that the sensor/transducer detects must be converted in to an electrical signal that is useful not only from interfacing point of view but also from noise point of view. Electrical signal may be voltage or current with single ended or differential ended connection.

As explained in Chapters 2 and 3, basic sensing parameters in chemical gas lasers are temperature, pressure and photo-signals. The sensors used in gas lasers require excitation current or voltage, bridge circuits, high amplification etc. for proper and accurate operation. This chapter deals with the signal conditioning required for sensors in chemical gas lasers. It includes signal sources and their types, operating voltage selection criteria, current sources and Wheatstone bridges for resistive sensors, amplifiers for small signal sensors, current to voltage converters for photo diodes.

4.1 Signal Sources

Most of the transducers or signal conditioners provide the signal in the form of voltage. When signal is transmitted over a long distance or is susceptible to noise then it may be converted in to a current signal prior to signal transmission. However, at the receiving end it is converted back in to voltage signal in most of the cases. The current signal is preferred for long distance transmission because the atmospheric noise tends to disturb voltage signal the most. Hence, current output ensures low susceptibility to noise. Therefore, thorough understanding of signal sources and their measurement methods is essential. The voltage signal source can be categorized in to two types:

- Grounded signal sources;

- Floating (ungrounded) signal sources.

4.1.1 Grounded Signal Sources

In the grounded signal sources, one of the signal leads is connected to the system ground as shown in Fig. 4.1. Thus, the voltage output from signal source is the potential difference between the positive signal lead of the signal source and the system ground. A typical example is instruments being normally earthed via its AC plug to the building ground. Electrical systems are grounded to protect circuits, equipment, and conductor enclosures from dangerous voltages and personnel from electrical shock. The grounded conductor carries the fault current back to the source and returns over the faulted phase and activates the overcurrent protection device.

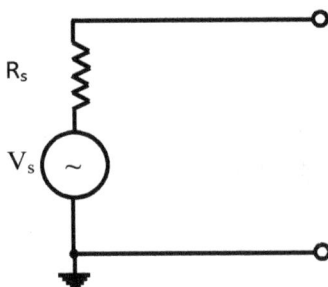

Fig. 4.1. Grounded signal source.

4.1.2 Floating Signal Sources

In floating or ungrounded signal sources, none of the lead of signal source is connected to the system ground. This means that the signal source is not referenced to any absolute reference. Fig. 4.2 shows the floating signal source. Examples of floating signal sources are transformer batteries and battery powered instruments. Ungrounded electrical systems are used where the designer does not want the overcurrent protection device to be activated in the event of a ground fault.

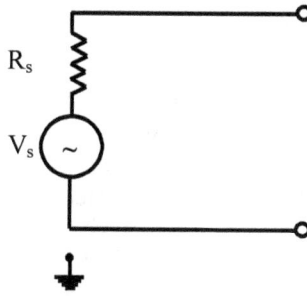

Fig. 4.2. Floated signal source.

All ungrounded systems should be equipped with ground detectors and proper maintenance should be carried out to avoid the overcurrent of a sustained ground fault on ungrounded systems.

The available data acquisition hardware provides options for measurement of signals generated from a source, which in present case is a sensor module. The measurement technique can be classified as:

- Single-ended;
- Differential-ended.

4.1.3 Single-ended Measurement

In single-ended measurement system as shown in Fig. 4.3, the voltage measurement is done with respect to 'ground'. It is called single-ended because only one signal line is needed to determine the signal voltage. Single-ended inputs are lower in cost, and provide twice the number of inputs for the same size wiring connector, since they require only one analog HIGH input per channel and one ground shared by all inputs. Hence, single-ended inputs save connector space, cost, and are easier to install.

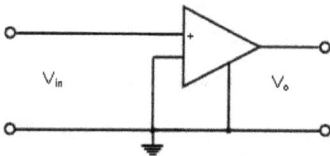

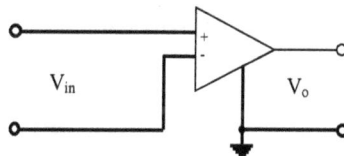

Fig. 4.3. Single-ended connection.   **Fig. 4.4.** Differential-ended connection.

4.1.4 Differential-ended Measurement

In differential-ended measurement as shown in Fig. 4.4, both the input leads are not linked to system ground. Differential measurement is advantageous because noise is induced equally as common mode voltage at both the input lines and is rejected.

Differential ended connections provide a more stable reading when Electromagnetic interference (EMI) or Radio frequency interference (RFI) is present. Therefore, it is recommended to use them whenever noise is generally a problem. This is especially true when measuring thermocouple inputs (in CO_2 GDL/HF-DF lasers), strain gauge and bridge type pressure sensors input (in CO_2 GDL/COIL/HF-DF lasers), since they produce very weak signals that are very susceptible to noise.

4.2 Analog and Digital Signals

The advent of computer technology and microprocessor based process control has a great influence on the development of digital transducers. Digital [3-4] signal have significant advantages over analog signals such as uniform format, speed, accuracy and compatibility with computers. However, analog signals represent real data and digitization of analog signal always results in quantization error that can never be recovered. But it can be made as low as possible by using more signal levels (number of bits) to represents an analog signal.

In real world, physical phenomenon, such as temperature, pressure and sound vary according to the law of nature and exhibit properties that change continuously in time. They are all analog time varying signals. Analog signals may have any value with respect to time. The relevant information contained in the signal is dependent on whether the magnitude of the analog signal is varying slowly or quickly with respect to time. The three basic characteristics of an analog signal are amplitude, shape and frequency.

A digital, or binary, signal cannot have any value with respect to time. Instead, it has only two possible specified levels or states: an 'on/ high' state, in which the signal is at its highest level, and an 'off/low' state, in which the signal is at its lowest level. Digital signals have certain specifications that define characteristics of the signal. For example, the output voltage signal of a transistor-to-transistor logic (TTL) switch can only have two states: the value in the 'on' state is 2.2 to 5.5 Volts, while the value in the 'off' state is 0 to 0.8 Volts. The useful information that can be measured from a digital signal includes the state and the rate.

4.2.1 Operating Voltage and Output Signal Selection

One of the most important factors in the design of any automatic/ electronics/sensor system is the selection of excitation voltage [5]. The chemical gas laser system incorporates various transducers, actuators and interfacing electronics circuits employing different integrated circuits (ICs). The main transducers are resistance temperature detectors/ thermocouples, pressure sensors, flow/level sensors and photo detectors etc. whereas main actuators are solenoid valves and electro-pneumatic valves. The design of digital output processing interfacing electronics circuitry and temperature to current converter circuitry is based on different ICs. All these devices require electrical power for their operation. Choice is available for their operation under different voltage conditions such as 5 Vdc, 12 Vdc, 24 Vdc, 48 Vdc, 110 Vac and 220 Vac etc. Also, the output from various transducers may be 0 to 5 V, 0 to 10 V, any non-zero voltage to 10 V, 0 to 20 mA, 0 to 5 mA, 4-20 mA etc. Thus design of DAS demands optimum selection of excitation voltage and sensor signal among the available options. The excitation voltage may be divided in to three groups:

- Low voltage (typically 5 Vdc or less);
- Medium DC voltage (12 Vdc, 24 Vdc, 48 Vdc);
- High voltage 110 Vac, 220 Vac.

The low voltages are very susceptible to receiving interference where high voltages are prone to causing interference. The medium DC voltages (specially 24 Vdc) tend to be in the middle of the two extremes. That (plus a few other reasons like current draw) is why every one has been "standardizing" on 24 Vdc for industrial applications. Thus, 24 Vdc can be selected as the operating voltage for entire operation of chemical gas laser's DAS.

The standard electrical signals are 0 to 100 mV, 0 to 10 V, -10 to +10 V, 0 to 20 mA, 4 to 20 mA etc. However, 0 to 10 V and 4 to 20 mA are most common type of signals used in industrial instruments and sensors and is termed as a 4-20 mA loop. The main advantage of using this signal is that the current is same in each part of the circuit in the loop and hence each component will receive the same signal. Another advantage is that it represents a *live zero*. 4 mA represents a minimum measured amount (which is often a zero value) of physical parameter whereas 20 mA represents the maximum measurand value. In 4-20 mA signal case, there is always 4 mA current in the loop and if it drops to zero then link break condition is detected and it may be used as a fault detection of link breakage. The current output is varied in this range depending upon the varying impedance created by the transducer. The current output can be converted in to voltage by inserting a resistance R_L into the loop as shown in Fig. 4.5.

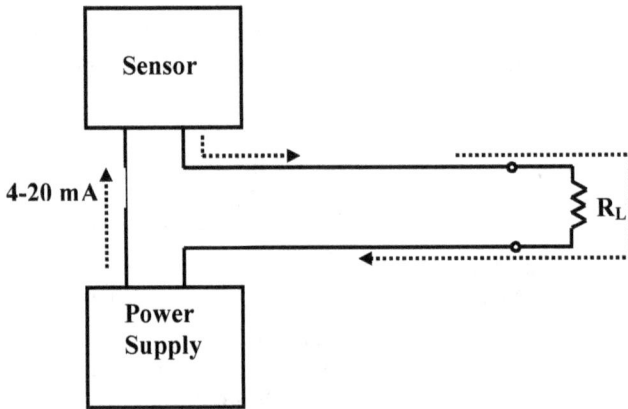

Fig. 4.5. 4-20 mA current loop.

Thus, the design of signal conditioning and interface circuits for chemical gas lasers should be based on the 24 Vdc supply and the selection of all transducers, actuators and other devices has to be done for this voltage. The transducers/ sensors should either be selected to produce 4-20 mA current output or they should be interfaced with signal conditioning module to produce 4-20 mA current output for acquisition by analog to digital converter.

4.3 Sensor Circuits for Signal Conditioning

Since most of the sensors e.g. thermocouple [6], provide a very small signal, it must be amplified or compared before it can be used. For signal amplification, operational amplifier based amplification circuit can be designed for example, inverting amplifier, non-inverting amplifier and instrumentation amplifier. Also some of the sensors works on the principle of change in resistance [7] (for example RTD) with respected to measurement parameter. The change in resistance can be measured either by using Wheatstone bridge or by using constant current source. The photo detector output is in current form, which requires current to voltage conversion circuit.

4.3.1 Wheatstone Bridge

Wheatstone bridge [8] circuit of Fig. 4.6 can be used for detection of change in resistance due to change in physical parameter. For example, RTD can be connected to one of the arms of Wheatstone bridge. The variable resistor in the bridge is adjusted to provide the set point. As the temperature changes, the resistance in the RTD will change accordingly and voltage through the bridge will reflect this change. Any resistive sensor can be used in bridge circuit. Strain gauge based pressure sensors also normally utilize this bridge circuit.

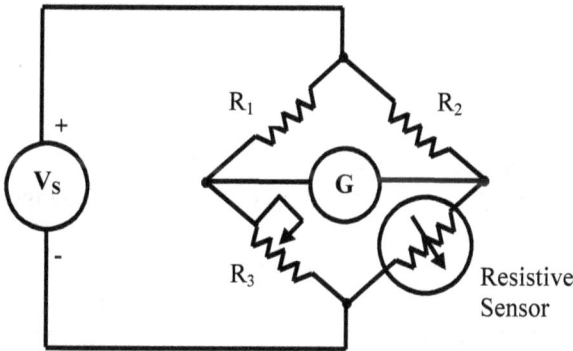

Fig. 4.6. Resistive Sensor connected in bridge circuit.

Measurement of resistance in RTD requires a small current to be passed through the sensing element. This can cause resistive heating (self-heating) which will affect the accuracy of the actual temperature measurement being made. Inaccuracies in temperature measurement due to self-heating can be greatly reduced by either minimizing the excitation power or exciting RTD only when measurement is required. Mechanical strain on the resistance thermometer can also cause inaccuracy. Lead wire resistance can also be a factor but its effect can be nullified using three or four-wire RTD configuration. Fig. 4.7 shows a circuit of Wheatstone bridge in which three wire RTD is used and lead resistance is nullified. During balance condition, there is no current to flow through lead resistance R_{L3} of RTD, whereas, lead resistance R_{L1} & R_{L2} are placed in those arms canceling the lead resistance effect.

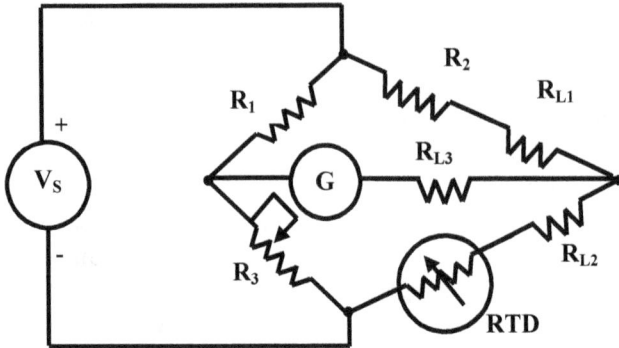

Fig. 4.7. Three wire RTD connected in bridge circuit.

4.3.2 Constant Current Sources

Constant current source may also be used to convert the change in resistance to voltage signal according to temperature change. We know that $V = I \times R$, since current I is kept constant; hence voltage (V) will vary due to change in resistance(R) with respect to temperature change. There are various configurations of basic current sources. Fig. 4.8 shows a constant current source using Zener diode in which collector current is given by the relation below, which is dependent on Zener voltage V_Z and base to emitter voltage (V_{BE}). V_Z of zener diode is constant and V_{BE} and α of a transistor is also constant. Hence, collector current I_C will be constant if R_E is constant.

$$I_c = \alpha \frac{(V_z - V_{BE})}{R_E} \qquad (4.1)$$

Similarly, Fig. 4.9 shows another basic configuration of constant current source using operational amplifier. The load current I_L is given by relation 4.2. If input voltage V_{IN} is constant then load current I_l will be constant for a given emitter R_E and supply voltage V_{CC}.

$$I_L = \alpha \frac{(V_{CC} - V_{IN})}{R_E} \qquad (4.2)$$

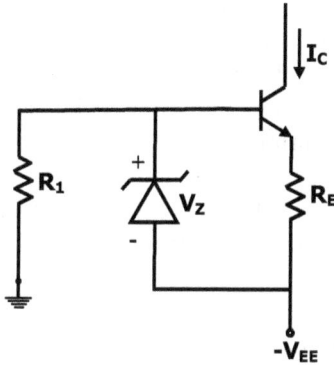

Fig. 4.8. Zener diode based constant current source.

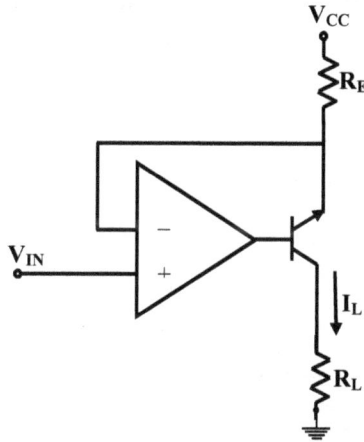

Fig. 4.9. Operational amplifier based constant current source.

Fig. 4.10 shows a Burr Brown IC XTR-105 [9] that may be used to convert change in resistance into 4-20 mA current output suitable for transmission over longer distances as required in high power chemical gas lasers. At the receiving end it is again converted into a voltage signal by employing a suitable load resistance.

This circuit has been used in chemical oxygen iodine laser for temperature measurement of iodine supply system (ambient room temperature to 150 °C) and basic hydrogen peroxide system (ambient room temperature to -25 °C). XTR-105 is a monolithic 4-20 mA, two-wire current transmitter. Two matched 0.8 mA current sources drive the RTD and zero setting resistors R_Z. The built-in amplifier input of the XTR-105 measures the voltage difference between RTD and R_Z. The value of R_Z is chosen to be equal to the resistance of the RTD at the minimum measurement temperature range. R_Z can be adjusted in such a way to get 4 mA current output at the minimum measurement temperature so as to correct for input offset voltage and reference current mismatch of XTR-105. R_{CM} provides an additional voltage drop to bias the inputs of the XTR-105 within their common-mode input range. In order to minimize the common mode noise, R_{CM} is bypassed with a 0.01 µF capacitor. Resistor R_G sets the gain of the instrumentation amplifier according to the desired temperature range. R_{LIN1} provides linearization correction to the RTD, typically achieving a 40:1 improvement in linearity. Load resistance R_L of 470 ohm can be taken to convert 4-20 mA current into the approximate voltage range of 0 to 10 V so that it matches with the dynamic range of analog input channel of data acquisition module (is discussed later in Chapter 5).

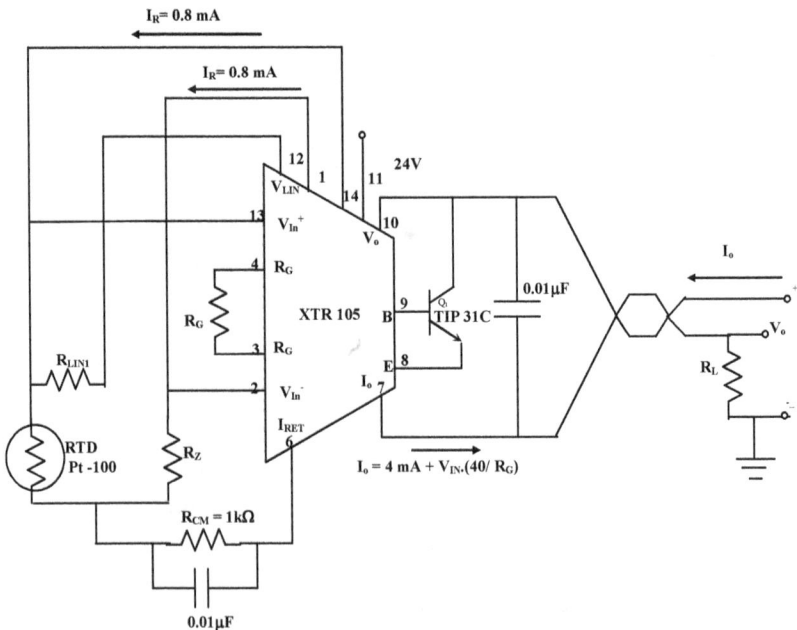

Fig. 4.10. Interface electronics circuit for RTD Pt-100 to produce 4-20 mA output.

4.3.3 Amplifier Circuits

The output of thermocouple is in the range of microvolt, which needs to be amplified. Amplification of signal is also required to match the output of sensor with dynamic range of analog to digital converter. A very simple amplifier may be based on inverting or non-inverting operational amplifier as shown in Fig. 4.11 and Fig. 4.12 respectively. However, these amplifiers will amplify common mode noise also.

For inverting amplifier configuration, the gain of amplifier (G_{Inv}) is given by relation (4.3), whereas, relation (4.4) shows amplifier gain ($G_{Non-Inv}$) in non-inverting configuration.

$$G_{Inv} = \frac{V_o}{V_{In}} = -\frac{R_F}{R_{In}} \tag{4.3}$$

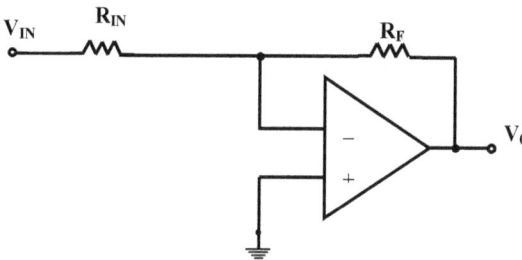

Fig. 4.11. Inverting amplifier circuit.

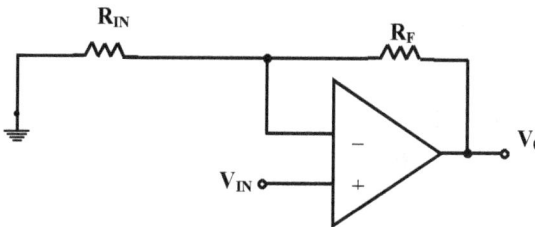

Fig. 4.12. Non-Inverting amplifier circuit.

$$G_{Non-Inv} = \frac{V_o}{V_{In}} = \left[1 + \frac{R_F}{R_{In}}\right] \tag{4.4}$$

143

Instrumentation amplifier [10] is commonly used for amplification of thermocouple output. Fig. 4.13 shows a typical circuit of instrumentation amplifier using three operational amplifiers.

Instrumentation amplifier circuit is a differential amplifier optimized for high input impedance and a high common mode rejection ratio (CMRR) and isolation between input-output. In this circuit, the inputs are voltage followers and they have very high input impedance, which effectively isolates the input from the output. The second stage is a differential amplifier, which subtracts the common noise present at both the inputs. The gain (G) of instrumentation amplifier is given by relation below:

$$G = \frac{V_o}{V_2 - V_1} = \left(1 + \frac{2R_2}{R_1}\right)\left(\frac{R_4}{R_3}\right) \qquad (4.5)$$

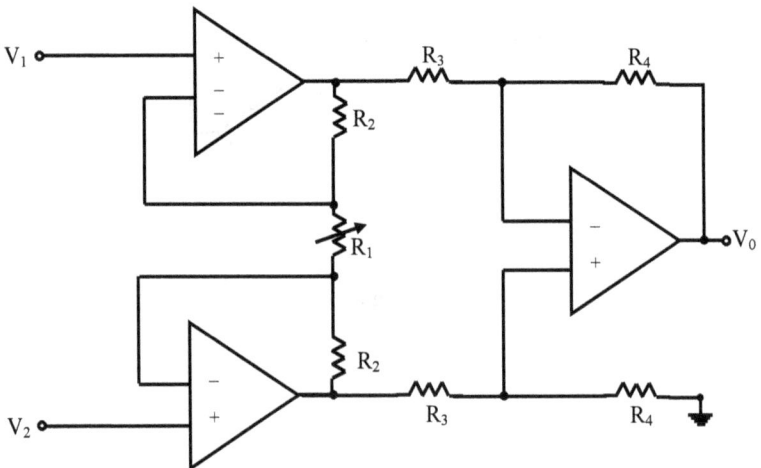

Fig. 4.13. Instrumentation amplifier for amplification.

4.3.4 Current to Voltage Converter

Photodiode produces a current with respect to the optical signal. Current to voltage converters are widely used for signal conversion in photodiode [11-12]. Photodiode can be operated with or without an applied reverse bias depending upon specific requirements. They are referred to as photoconductive (biased) and photovoltaic (unbiased) modes. Application of a reversed bias can greatly improve the speed of response and linearity of the devices. The reverse bias serves to accelerate the electron/hole transition times and improves the frequency response. Photoconductive diodes operate over a frequency range from dc to100 MHz and capable of measuring the intensity of light with the rise times in the range of 0.1 to few ns. Applying a reverse bias, however, will increase the dark and noise current. The photovoltaic mode of operation is preferred when photodiode is used in low frequency applications (up to 350 kHz). The semiconductor photodiodes are small, rugged and inexpensive and because of these advantages they have replaced the vacuum-tube detectors in many applications.

The signal conditioning circuits for twin modes of Photodiode are shown in Fig. 4.14 and Fig. 4.15 respectively.

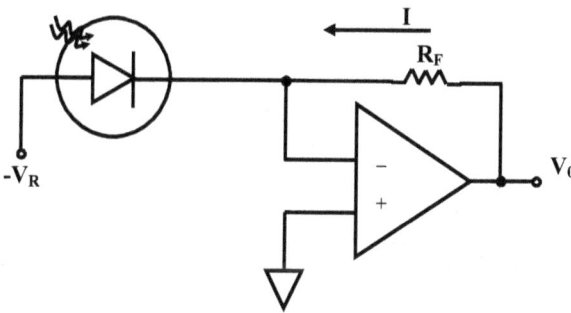

Fig. 4.14. Circuit for photodiode in photoconductive mode.

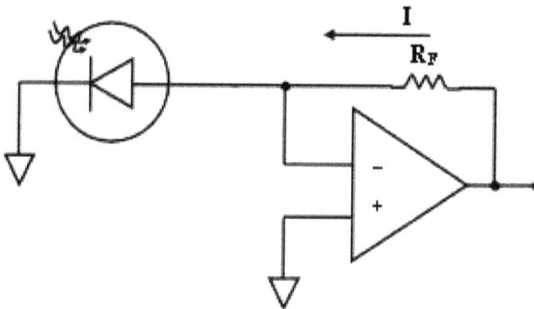

Fig. 4.15. Circuit for photodiode in photovoltaic mode.

References

[1] Khajan, Alexander D., Transducers and their elements, Eaglewood Cliffs, *Prentice Hall Publishing*, NJ, 1994.

[2] Allocca, John A., and Allen Stuart, Transducers Theory & Applications, *Reston Publishing, Inc.,* a Printice Hall Company, Reston, VA, 1984.

[3] Nihal Kulratna, Digital and analogue instrumentation: testing and measurement, *The Institution of Electrical Engineers,* London, 2003.

[4] Roger L. Tokheim, Digital Electronics, Third Edition, *McGraw-Hill Publishing Company,* 1990.

[5] Mainuddin, Gaurav Singhal, R. K. Tyagi, A. K. Maini, Diagnostics and data acquisition for chemical oxygen iodine laser, *IEEE Transactions on Instrumentation and Measurement*, 2012, Vol. 64, p. 1747.

[6] Baker, H. D., E. A. Ryder and N. H. Baker, Temperature Measurement in Engineering, Vols. 1 and 2, *Wiley,* New York, 1953, 1961.

[7] James R. Carstens, Electrical Sensors and Transducers, *Regents/Prentice Hall, Eaglewood Cliff,* N. J. 1993.

[8] Thomas E. Kissell, Industrial electronics: Applications for programmable controllers, instrumentation and process control, and electrical machines and motor controls, *Prentice Hall, Inc.,* 2000.

[9] www.burr-brown.com/databook/XTR105.htm.

[10] Patrick H. Garrett, High Performance Instrumentation and Automation, CRC Press, *Taylor & Francis Group,* FL, 2005.

[11] Govind P. Agrawal, Fiber-optic communication systems, Chapter IV, *John Wiley & Sons, Inc.,* 1992.

[12] M. Razeghi, Long wavelength infrared detectors, *Gordon and Breach Science Publishers,* 1996.

Chapter 5

Data Acquisition System and Safety Measures

Data acquisition system (DAS) plays a major role in the development of chemical gas lasers. Both research and development activity in this field demands a special custom-built data acquisition system. DAS should not only be able to handle online operational requirements but also capable of being interfaced with special diagnostics systems already discussed in Chapter 3. The operation time of these lasers is typically limited to a small duration for various reasons, hence, the supply and control of these gases has to be performed in rapid and precise manner.

The performance of gas chemical lasers is dependent on various phenomena inside the laser gas dynamic tunnel. Therefore, it should incorporate an analysis module for post run performance analysis for serving the purpose of trouble shooting, optimization and also for developing thorough understanding of these lasers. This requires on line monitoring, estimation and display of various parameters with high temporal resolution. Further, involvement of hazardous chemicals and gases requires safety interlocks for human safety. Thus, a dedicated and a highly versatile data acquisition system (DAS) is required to perform all these functions. As a step towards this, it is essential to develop a highly adaptable system, incorporating a fair magnitude of flexibility from the viewpoints of functionality, parameter monitoring, acquisition, diagnostics, safety interlocks and performance analysis. In addition to be able to fulfill these features, DAS should be capable of being operated by a single person enlaced with user's friendly features in form of necessary graphical user interfaces.

Thus, this chapter is focused on necessity, various aspects of design and development of multipurpose real time data acquisition system and its implementation for performance analysis of gas lasers.

5.1 DAS Requirements

Prior to going into the specifics of DAS design it would be worthwhile to comprehend and discuss objectively the main requirements of dedicated DAS for operation of a chemical gas lasers. The same are enunciated below,

- Precise control of gases flow e.g. Nitrogen, Iodine and Chlorine, time sequencing of these gases and liquid reagent flow (i.e. BHP in COIL and toluene in CO_2 GDL). The sequential operation in turn requires precision control of actuators associated with these processes.

- Requirement of changing the flow parameters during the operation, which is typically of the order of few seconds.

- Remote handling of toxic and hazardous chemicals like BHP solution, chlorine, iodine, hydrogen, toluene etc. It requires suitable interlocks and also interfacing of safety equipments in case of any failure.

- Online display of acquired parameters such as temperature, pressure, flow during laser preparation and storage of all measured parameters during laser run for post run performance analysis and optimization of gas lasers.

- Diagnostics system implementation in DAS for determination of relevant parameters such as Mach number, small signal gain, specie concentration, medium homogeneity etc.

- Design of user's friendly bitmap control windows (graphical user interfaces) for ease of operation by a single operator.

Keeping in view these requirements, it is important that a research grade chemical gas laser should be operated with a dedicated data acquisition system (DAS), where all the controls are in the hands of a single operator.

5.2 Computer Based DAS: Overview

There has been tremendous progress in the field of computer based data acquisition systems (DAS) during the last decade. DAS typically involves acquisition and storage of signals from various sensors, generation of signals to control the processes and presenting them on graphical user interface for the display and parameter analysis. The main attractions behind the popularity of computer based technology are low cost, accuracy, flexibility and simplicity of implementation. Further, flexibility and functionality allows development of DAS in a variety of ways each having distinct features depending on the applications envisaged. Moreover, 'off-the-shelf' plug and play components with menu driven software packages are available, which greatly simplify the task of a designer in terms of development of customized DAS.

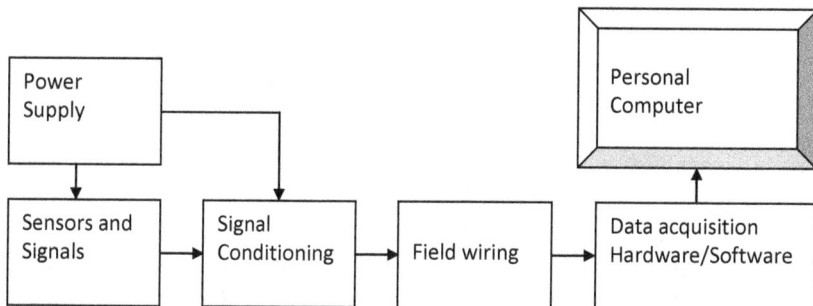

Fig. 5.1. A typical computer based data acquisition system.

A PC based DAS, generic schematic is shown in Fig. 5.1, consists of following basic building blocks:

- Transducers / sensors;
- Signals;
- Signal Conditioning;
- Field wiring;
- Data acquisition hardware;
- Data acquisition software;
- Personal computer.

A simplified block diagram of a computer based data acquisition system is shown in Fig. 5.2.

Data acquisition has the function of primarily gathering information about a system or process. It is a core tool for control, management and developing a lucid understanding of a system or a process. Basic parameter information such as temperature, pressure or flow is gathered by the sensors, which convert information into an electrical signal. In certain cases one sensor may suffice, such as while monitoring local temperature. However, in many cases even hundreds or thousands of sensors may be required, say for monitoring a complex industrial process or a complex chemical gas laser. Thus, development of DAS starts with the *transducers/sensors,* which produce the electrical signal with respect to the change in physical parameters e.g., temperature sensors, pressure sensors, flow meters, photo detectors etc. The signal from these transducers may or may not be suitable from control and acquisition point of view; and hence signal conditioning is required for this purpose, which has already been discussed in Chapter 4. Subsequent to signal conversion, sensors are connected to the data processing cards via *field wiring.* The *data processing cards* not only perform the task of analog to digital and digital to analog conversion but also the acquisition and generation of digital signals. The *interfacing* of cards with computer is another major concern. The card may be connected to a computer in a number of ways such as; serial port, parallel port or system expansion bus. *Software* is as important to data acquisition systems as the hardware capabilities. Inefficient software can waste the usefulness of the most able data acquisition hardware system. Data acquisition software controls not only the collection of data, but also its analysis and eventual display. Ease of data analysis and presentation are the major reasons behind using PC for data acquisition in the first place.

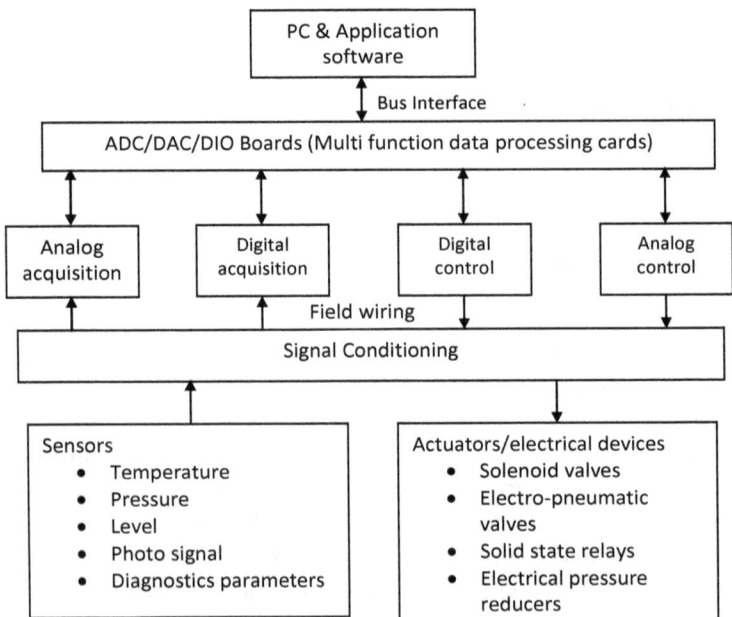

Fig. 5.2. Simplified block diagram of computer based DAS.

The typical DAS related requirements of chemical gas lasers along with possible options are shown in Table (5.1). The first step in the design of DAS is the selection of basic approach, which can be based on either Personal Computer or Programmable Logic Controller. The second step in the design of DAS is ensuring its compatibility with set of selected sensors, which are governed by the application. The selection of sensors for chemical gas lasers has been discussed in great detail in Chapter 2. Table (5.1) repeats some of these aspects for the sake of completeness. Also, as DAS serves the operational requirements of the system it should be capable of working in conjunction with suitably selected actuator/ electro-pneumatic devices. The major factors deciding the kind of actuators are response time, nature of fluid and supply voltage. The selected appropriate actuators and sensors are to be interfaced suitably with data acquisition hardware. The selection of data acquisition hardware includes analog to digital converter, digital to analog converters and digital input/output cards. The software package/application software selection is also very important, which finally results in complete data acquisition system. The field wiring aspects also need proper attention by making careful selection of signal wiring cables and connectors.

Table 5.1. DACS related Requirements of Chemical gas lasers with possible options .

S. No.	Requirements	Governing Parameters /Functions/ Devices	Salient features /Ranges	Possible options
1	Control, acquisition and analysis functions	Compact and flexible DAS	100-200 channels	PC/PLC based technology
2	Operating voltage	Power supply	DC/AC	5/12/24/48 V dc 110/220 V ac
3	Sensors	Temperature	$10°C$ to $200°C$, $-50°C$ to $+50°C$, $10°C$ to $1800°C$	Thermocouple/RTD/ Thermistor/ICs
		Pressures	0 to 100 torr & 0 to 1000 torr, 0 to 100 bars and 0 to 200bars	Capacitance gauge/ diaphragm type
		Flow	Few mmol to few hundred mmol/sec	Orifice under choked conditions /rotameter/ turbine flow meters

Table 5.1. (Continued).

S. No.	Requirements	Governing Parameters /Functions/ Devices	Salient features /Ranges	Possible options
3	Sensors	Optical signal	~499 nm, 330 nm, 1315 nm, 2.7µm, 3.7 µm, 10.6 µm	Silicon/Germanium photo diodes /Indium antimonide detector/Mercury cadmium telluride detector
4	Actuators/ electro-pneumatic components	On/off control of buffer gases/others	Operating temp. room & pressure ~10 bar	Solenoid valves
		On/off control of iodine and BHP supply	Operating temp. ~+70° C & ~20° C, corrosive medium	Electro-pneumatic valves with special sealing
		Flow rate control	Control pressure 1 to 9 bar with analog output for pressure monitoring	Electrical pressure reducers
5	Data acquisition hardware	Analog to Digital Conversion	Few hundred samples/sec, 16-bit resolution & 0 to 10 V input signal level	Dual slope/ Successive approximation/ Flash type ADCs
		Digital to analog conversion	0 to 10 V analog output voltage with ~2 mA driving capability	Weighted resistor/ R-2R resistor Type DAC
		Digital input/output boards	5 V TTL signal	DIO boards with TTL signal compatibility
6	Data acquisition software package	Application software development	Graphical tools & suitability with DAQ hardware	GeniDAQ/DasyLab/LabVIEW
7	Signal cables/connectors	Wiring layout and connections with control panel	Shielded cables along with connectors	Commercially available standard cables/connectors

5.3 Design Methodology of PC based DAS

There are two main approaches for the development of data acquisition and control systems. The first one is based on programmable logic controllers (PLC) technology [1-4], whereas second one is personal computer (PC) based technology [5-11]. It has always been an issue of debate that which one is better. Both of them have certain advantages and disadvantages. In 1960's the first PLC was introduced to provide a substitute for the large number of electromechanical relays that were being used in industrial control circuits. Earlier efforts used for automation employed large numbers of relays and their wiring was quite complex. Typically, a machine control panel includes possibly hundreds or thousands of individual relays. As a result any change in design in the control circuits was quite time consuming. Moreover, the relays not only had a limited lifetime but also had a slow response. The Modular Digital Controller (Modicon) [12] brought the world's first PLC into commercial market. Several major companies such as Allen-Bradley, Texas Instruments and Siemens etc. are the main manufacturers of PLCs. PLCs allowed a program to be written and stored in memory, and as and when changes were required to the control circuit, changes could be made in the program rather than in the electrical wiring. It also became feasible for the first time to design one machine to do multiple tasks by simply changing the program in its controller.

The challenges of machine and process control system designs have for many years been addressed through PLC route. This approach has been based on the installation of a programmable logic controller from a leading vendor such as Allen-Bradley [2], Siemens [13], or Cutler Hammer [14]. Such PLCs are usually proprietary systems implying that when one selects a particular vendor and PLC family, then his applications are restricted to the corresponding boards and functions that are available to that particular line. It is not possible to modify them to a large extent from standard configurations. The communication interface, statistical functions, data acquisition functions and new sensors are often difficult to interface with traditional PLC hardware. Although, the PLC technology enjoys the advantages of reliability and ruggedness but the main drawback with PLC is that they have dedicated proprietary operating and software systems. An Allen-Bradley PLC program does not run on a Siemens PLC and vice versa. Because of these challenges, the PLC is finding only limited users now. On the other hand, the functionality, flexibility and compatibility are the major advantages of computer technology over PLC technology.

The main attractions behind the popularity of computer based technology are low cost, high flexibility, relatively simple to implement and high performance. 'Off-the-shelf' components are available which can be interfaced with computer easily. The computer technology along with the software based operator interface has got attraction in the applications such as research and development activities in laboratory, process control in industries, and automatic test facility for inspection of hardware. It also finds applications in medical diagnostics, chemical and hazardous industries and environmental diagnostics and control. The compatibility and standardization of plug and play hardware with menu driven software package has simplified the realization of custom based application with minimum efforts.

The functionality and flexibility of computer allows the DAS to be configured in a number of manners, each with its own distinct advantages. The efficient designing of data acquisition systems along with the effective use of PC involves the careful matching of the specific requirements of a particular data acquisition application to the appropriate hardware and software available. Table (5.2) shows the summary of comparison of PLC and PC based technology.

Table 5.2. Comparison between PLC and PC technology.

S. No.	Features	PLC	PC
1	Functionality	Limited	Enormous
2	Flexibility	Limited	Highly flexible
3	Architecture	Closed	Open
4	Addition of new feature	Difficult	Very Easy
5	Software compatibility	Proprietary	Very good
6	Plug and play hardware	Not possible	Available
7	Programming	Difficult	Very easy
8	Ruggedness	Higher	High
9	Reliability	Higher	High
10	Cost	Costlier	Most economical

As discussed above, personal computer based approach enjoys a lot of advantages over PLC approach, hence, personal computer based technology is best suited for development of DAS for chemical gas lasers.

The present session on design of DAS is discussed in backward step process starting from the laser system moving towards the data acquisition cards describing the manner in which the signal is processed.

5.3.1 Operating Voltage

The topic has already been discussed in Chapter 4, in which we had concluded that design of DAS for chemical gas lasers should be based on the 24 V dc supply and the selection of all transducers, actuators and other devices has to be done accordingly. The transducers/ sensors should either be selected to produce 4-20 mA current output or they should be interfaced with signal conditioning module to produce 4-20 mA current output for acquisition by analog to digital converter.

The selection of sensors has been discussed in detail in Chapter 2 but it would be prudent say a few words with reference to actuators. The actuators have a very important role in the operation of chemical gas lasers. The sequential on/off control of the supply of various chemicals and gases needs to be performed using various electrical and electro-pneumatic actuators. Since chemical gas lasers deal with not only nitrogen gas but also hazardous medium like chlorine, iodine, fluorine, and BHP/ Toluene solution. Hence the actuators selection for N_2 and corrosive gases will be different. For N_2 solenoid valves, (SMC/Rotork) may be selected whereas for chlorine, iodine, hydrogen BHP/ Toluene solution supply electro –pneumatic valves (Burkert, /BiTorq) may be selected. These valves are made of SS-316 with PTFE sealing; thus making them suitable for aggressive fluid as in chemical gas lasers. The maximum operating pressure of this valve is 16 bar and minimum control pressure is 2.5 bar that is controlled by a small solenoid valve fitted with the body of valve itself. Again the operating supply voltage of both of these valves is 24 V dc as per system requirement.

5.3.2 Interfacing Circuits

DAS systems are used in wide range of applications that employ various types of transducers. Sometimes transducers generate signals that are not suitable to be measured directly with a DAS device. Thus, signal from many transducers must be conditioned in some way before a measuring system can accurately acquire the desired signal. The signal conditioning for sensors at the sensor end applicable for chemical gas lasers has already been detailed in Chapter 4.

Also, in chemical gas laser applications appropriate electronics circuits are required for interfacing data acquisition hardware with actuators or similar other electrical devices at the DAS end as well. The multi function DAS card generally meets most of the requirements of DAS for gas laser operation. Digital input of the DAS card can be directly linked with the subsystems. Similarly the analog input of the DAS card can accept 4-20 mA or 0 to10 V signal from the subsystems directly. The current driving capability of analog output of the used DAS card is ± 20 mA. It can directly run say an electrical pressure reducer, used for operating a system at fixed pressure, which requires only ~1.5 mA.

One of the areas of concern as far as multi function DAS card is the digital output for which Digital output interface processing electronics circuit is required at DAS end. Usually, it is not possible to drive electrical devices directly through Digital output (5V TTL) of data acquisition card because of the current limitation. Further some of the devices like solenoid valves and electrical pressure reducers operate at 24 V dc. The interface electronics circuitry [15] is designed for digital output (5 V TTL) keeping in view the requirements of data acquisition card. Typical scheme that may be used for the DO interface-processing unit is shown in Fig. 5.3.

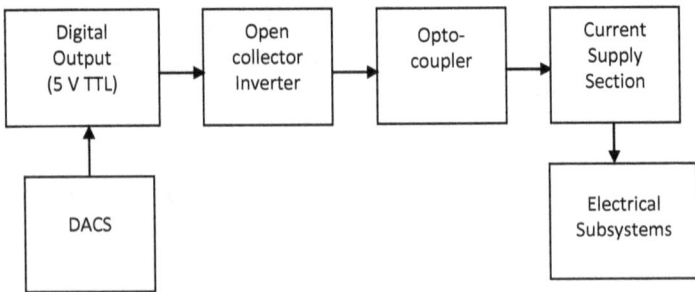

Fig. 5.3. Scheme used in the interface electronics for digital output.

The detailed electronics circuit diagram and a typical developed hardware for the digital output for our chemical laser module is shown in Fig. 5.4 and Fig. 5.5 respectively. The DO is fed to open collector inverter IC7406; output of this IC is connected to an opto-coupler 4N35. The optical de-coupling is required to isolate the data acquisition card from the electrical devices in the circuit. The current supply section employs the transistor TIP31C to meet the current requirements of various electrical devices. For example, in case of heater, which requires 1-10 A at 220 V ac, transistor TIP31C output (24 V dc) can be used to control the solid-state relay (SSR), which in turn drives the heater.

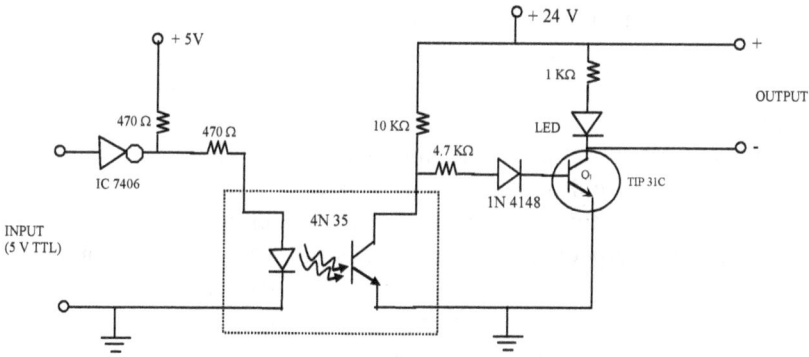

Fig. 5.4. Electronics circuit diagram for digital output.

Fig. 5.5. 16-channel DO processing interface electronics hardware with optical de-coupling.

In a single board 16 channels are implemented so that it can be directly connected with the PCLD-8710 screw terminal board through FRC cable. Four such DO electronics module were developed to fullfil the requirements of 64 DO channels. This circuit is directly used to drive all the electrical/electro-pneumatic valves which operate at 24 V dc. However for the devices operating at 220 V ac, this circuit is used in conjunction with solid state relays with different current ratings as per the load requirements.

5.3.3 Signal Cables and Connectors

Cable wiring is one of the most important tasks, which needs dedicated attention from data acquisition personnel. This passive component of DAS system is often overlooked as an important integral part. This may make the reliable system inaccurate and unreliable due to incorrect wiring techniques. *Field wiring* represents the physical connection from the transducers to signal conditioning hardware and/or data acquisition hardware. The field wiring often physically represents the largest component of the total system; it is more susceptible to the effects of external noise, especially in harsh industrial and field environments. The correct earthing and shielding of field wires and communication cabling is of great importance in reducing the effects of noise. The wiring does not simply mean to connect the wire between the transducer leads and signal conditioners or DAS device, or connecting the signal conditioners to DAS device itself. There are various methods for taking measurements of input signals by signal conditioner/DAS hardware. In order to have accurate and error free measurements, careful attention must be given not only to the type of signal produced by the transducer but also to the signal cables and signal connectors.

The high quality shielded signal cables can be selected for wiring layout of the DAS of chemical gas lasers. The screened signal cable from M/s RS component stock no. 205-1252 or equivalent may be used which employs 4 cores; 7/0.1 mm tinned copper stranded conductors with 0.2 mm PVC sheath. It is again surrounded by tinned copper braided screen along with outer grey PVC sheath. This cable has a maximum working voltage of 250 Vrms with maximum current pre core of 0.25 Arms. The core/screen capacitance is 85 pF/m and resistance per core is 384 mΩ/km only. The connector selection is also as important as the cable selection. Hence miniature circular connector series 723 from M/s Binder (Germany) can be selected for making connections at control panel. The 4 pin male socket and female plug has contact resistance ≤ 5 mΩ with insulation resistance $\geq 10^{10}$ Ω and the rated voltage is 250 V. This connector is capable of operating in the temperature range of $-30°$ C to $+95°$ C. These are just examples to illustrate to the reader the manner in which cable and connector may be selected. Although, reader would appreciate that the selection of these is primarily dependent on DAS hardware, sensor type which are all application dependent.

5.3.4 Data Acquisition Cards

The data acquisition hardware is basically the interface between the computer and the system. It primarily performs the following basic functions:

- The processing and conversion of analog input signals (measured from a system or process) in to digital form using Analog to Digital converters (ADCs). The digital data is then transferred to the computer for further processing, display, storage and analysis.

- Acquisition of digital input signals which exhibit the binary information about a system or process (e.g. system readiness in terms of yes/ no or open/close status of a valve).

- The processing and conversion of digital signals (from the computer) in to analog form using Digital to Analog converters (DACs). The analog control signal can then be used to control a process or a system (e.g. electronic pressure reducer requires analog signal as discussed in Chapter 3).

- The generation of digital control signals which can be used for performing switching and sequential on/off control of various actuators and systems.

In order to implement all above four functions, DAS [16-17] needs to have analog input channels, digital input channels, analog output channels and digital output channels along with counter/timer. The DAS boards are available from all the leading companies with one or all types of channels as per application requirements.

In order to develop data acquisition system, the choice of hardware and software is highly dependent on the specific requirements of the application and the environment in which system will operate. The number of channels (analog as well as digital), speed and resolution are the main factors, which decides the type and number of

ADC/DAC/DIO boards). However, multifunction card are available which provide all type of channels (analog input channels, analog output channels, digital input channels and digital output channels) on a single board with required sampling rates and resolution and bus interface (PCI/ISA).

The important manufactures of DAS hardware and software are National Instruments, Keithley Instruments Inc., Data Translation Inc., Scientific Solutions Inc., Gage Applied Inc., Microstar Laboratories, Omega and Advantech. National Instrument's LABVIEW and Advantech's GeniDAQ software packages along with a variety of hardware are most commonly used data acquisition products which show the compatibility [18] with all common high level languages. The various techniques of ADCs, DACs and DI/O boards are discussed in the forthcoming sessions:

5.3.4.1 Analog to Digital Converters

The ADCs [19-20] take the analog input signal and perform three basic functions: sampling, quantizing and encoding. During the late 1980s, much architecture for A/D conversion was implemented. Some of the important methods are *dual slope integration, successive approximation and flash type* analog to digital conversion.

Dual slope integration ADC uses an indirect method of A/D conversion, which relies on integration. The analog input voltage is converted to a time period that is measured by a counter. This is a very popular architecture in applications where a very slow conversion rate is acceptable. A classic example is the digital multimeters. This technique is very accurate because the value of the capacitor, resistor and conversion clock does not affect accuracy as they act equivalently on the up-slope and down-slope. The ratio of the charge time to the discharge time is equal to the ratio of the reference voltage to the unknown voltage. The absolute values of the resistor, capacitor and the clock frequency therefore do not affect the conversion accuracy. Extremely high-resolution measurement can be obtained for slowly varying signals with comparatively low cost. The main disadvantage of this technique is its slow conversion speed. The typical conversion speed of this type of ADC is few tens of samples/sec. This technique is mostly used in low frequency applications such as temperature measurements and digital multimeters.

The successive approximation ADC is the most common and popular direct A/D conversion technique used in data acquisition systems because it offers high sampling rates and high resolution with resonable cost. The main advantage of this method is that its conversion time is fixed and directly proportional to the number of bits of digital output. In this method, the addition of each successive bit doubles the ADCs accuracy whereas the conversion time increases only by the approximation period (T) of a bit. This technique is very simple and gives reasonably fast conversion rate. The 12-bit SAR ADC with conversion rate of upto 1 MSamples/sec and 16-bit SAR ADC with 250 KSamples/sec are commonly available.

The fastest type of ADC is *Flash converter* which is also known as *parallel* ADC. This type of ADCs are used when the extremely high speed conversion rate is required such as general high frequency applications and digital signal processing applications etc.Flash type A/D conversion is faster than the other methods because each bit of digital output code is produced simultaneously, irrespective of the number of bit resolution. It has a conversion rate ranging from MSamples/sec to over GSamples/sec. However, for the higher resolution, larger numbers of comparators are required to perform conversion. The addition of each bit doubles the number of comparators and resistors, therefore size and cost of the chip increases. Hence, in applications requiring little resolution and great speed, it is very useful.

Incase of chemical gas lasers, successive approximation is the best suited since it provides sufficiently fast conversion rates along with high resolution.

5.3.4.2 Digital to Analog Converters

Digital to analog conversion [8] is an essential function in the data processing systems. The digital to analog converters (DAC) provide an interface between the digital domain (host computer) and the analog world. The DAC converts the digital signal of the computer in to analog form that can be utilized to control the devices like various actuators or can be used to simulate a system or process. The digital to analog converter accepts an n-bit parallel digital data as input and produces an analog output signal with the help of summing resistive network which sums the current as per the binary value at the input.

The summing networks are mainly of two types: *weighted resistor network and ladder resistor network.*

In the weighted resistor form, the resistor values are taken in the bit weight ratio (1:2:4:8:16 and so on). The main advantage of this type of configuration is that only one switch and one resistor is needed per bit. It has the disadvantage of requiring many different values of very low tolerance resistors. R-2R resistor method overcomes this disadvantage of requirement of different values of resistors in weight resistor method because in this method, only two different values of resistors are required. Therefore, it is most appropriate to be used in chemical laser systems.

5.3.4.3 Digital Input/Output Boards

Apart from digital to analog converters and analog to digital converters, another important DAS hardware is Digital input and output (DIO) devices that find large application in any PC based DAS system. Digital I/O devices are mainly used to provide monitoring and control of industrial processes e.g. digital signal sensing, switch status monitoring, power/signal switching, device on/off control, valve/solenoid control and activating alarms etc.

Normally, DIO devices consists of ICs capable of accepting and producing TTL-compatible signals in which 0 to 0.8 V signal is treated as low level logic and 2.2 to 5.5 V is considered as high level logic. The important parameters of DIO board are number of digital input and output lines, the rate at which the data can be accepted or generated on these lines and the driving capability of the lines. If the digital output lines are used to control the heaters or lamps, a high data rate is usually not required because these devices cannot respond fast. In this application, the most important consideration is that the current requirement of the heaters/lamps (may be 5 A/220 Vac) must be less than the driving capability of DIO boards. Thus the interface electronics circuitry is required which can take TTL input and can meet current/voltage requirements of the heaters/lamps.

5.3.4.4 Interfacing

The communication interface between DAS board and PC is a very important aspect that needs attention. The DAS devices may be connected with PC via serial port or parallel port or it can be connected directly through system expansion bus. The parallel communication is faster than serial communication. But parallel communication interface has its own drawbacks. The long cable length, in parallel interface, increases the capacitive coupling between adjacent signal lines, producing cross-talk errors. Thus, parallel interfaces have cable length limitations, normally few meters. Whereas, serial interface uses few active signal wires, thereby, cross talk is minimized. Since fewer wires are required in serial interface, the cable is less expensive than parallel one. The common serial interfaces are RS232C, RS485 and universal system bus (USB) whereas general-purpose interface bus (GPIB) and Small Computer System Interface (SCSI) are important parallel interfaces.

The serial communication interface is the most widely used interface because it offers low cost cabling with potentially long cable lengths. RS232C can provide reliable communication up to 15 m with possible data rates of about 20 Kbps. From data acquisition and control application point of view, the RS232 interface has many limitations. The point-to-point restriction is the main limitation because in DAS application several devices are needed to communicate to each other. Secondly, this standard cannot be used for more than 15 m, which is a very short distance from DAS angle. Thirdly, the maximum data rate 19.2 Kbps is too slow for many DAS applications. Another popular serial transmission interface standard is RS485, which offers the maximum data rate of about 10 Mbps with the cable length of 1200 m. The multi-drop feature of RS485 is very useful in data acquisition applications because many DAS modules may be connected together on the same line.

In order to keep pace with advances in PC speed and performance, newer serial interface standards have been developed in which namely universal serial bus (USB) is very famous. The USB is a high speed and multi-drop serial bus that is developed to replace the standard serial and parallel ports on PC. Small DAS have traditionally suffered from the need of an easy to use and standardized bus system for connecting DAS devices. With its plug and play ability, it is extremely easy to implement and use. It is now a standard on all IBM compatible PCs. The USB standard can provide data rates up to 480 Mbps. In addition, it can provide DC power to peripheral devices (5 V at up to 5 A) and is hot swappable. However, the biggest problem with this standard is its maximum cable length (5 m), which restricts its use in DAS applications.

The IBM personal computers enjoyed great success because it adopted open system architecture design. This open system design permits the development of add-on products by the other manufacturers. This allows the DAS manufacturers to develop the DAS hardware that can directly be connected on the system bus. The commonly available expansion bus standards are industry standard architecture (ISA), extended or enhanced industry standard architecture (EISA), personal or peripheral computer interface (PCI) etc. The ISA expansion bus, initially resulted in 8 bit bus architecture with 4.77 MHz clock frequency. With the advent of 80286 16-bit processor, 8-bit ISA was extended to the 16-bit ISA with 6/8 MHz speed. With the introduction of Intel 32-bit 80386, IBM devised 16/32-bit bus with 10 MHz speed, known as microchannel architecture (MCA). But this bus was not compatible with ISA bus expansion boards at all. Hence, other manufacturers produced 386-based ISA machines. In 1988, they introduced a new standard extended industry standard architecture (EISA) in direct opposition to IBM's MCA bus. EISA is a 32-bit bus with 8 MHz speed because of the ISA compatibility restriction. But by doubling the data bus to 32-bit, it doubled I/O throughput. However, EISA never became very popular because of its relatively high cost.

In the early 1990's, VESA local (VL) bus was developed to improve the video based applications for better performance by Video Electronics Standard Association (VESA). VL bus was 32-bit wide with the speed of up to 50 MHz. But soon it was replaced by PCI bus, developed by Intel as a processor-independent bus. It was initially developed with 32-bit bus and 33 MHz speed. Its later versions support 64-bit bus with speeds of 66 MHz. The PCI-X is a high performance extension to the PCI bus that doubles the maximum clock frequency to 133 MHz with 64-bit transfer. The PCI bus is relatively new device to the PC motherboard. In near future, PC's will have more and more PCI slots and less and less ISA or EISA slots because of several advantages of PCI technology over ISA. The ISA or EISA were suited when only one application is running at a time. But now a days PC can run more than one application at a time. Secondly, card size of PCI bus is much smaller with more surface mount components to accommodate more functions at reduced cost. Moreover, PCI bus is a better plug and play device whereas in case of ISA/EISA recognition plug and play was difficult. PCI is a robust interconnects mechanism designed specifically to accommodate multiple high performance peripherals for graphics and video based applications. Therefore, PCI has significant advantages over ISA bus interface, hence PCI based approach is the best choice for DAS of chemical gas lasers.

For chemical gas laser application, the typical requirement of the acquisition speed of data is in the range of few hundred samples only; hence successive approximation technique is best suited in this application. The dual slope is very slow can support only few tens of sample per second and hence does not fulfill application requirements. Whereas, the flash type technique is very fast (of the order of Msamples/sec), very expensive with low resolution, hence it is not a wise decision to use this technique. For digital to analog conversion, R-2R resistor ladder technique is the better choice because of the simplified requirement of weighted resistances. Keeping in mind the other requirements of resolution and conversion time and accuracy, the PCI bus based Advantech PCI-1716 multifunction card may be selected as DAS hardware for ADC/DAC/DIO functions. The PCI bus selection is based on the fact that it is high speed and high performance bus as compared to ISA bus (max speed only 8 MHz) and moreover it is small in size and provides a processor-independent data path between the CPU and high-speed peripherals.

5.3.5 Data Acquisition Software

Data acquisition hardware does not work without software because software transforms the PC and DAS hardware in to a complete data acquisition, control, analysis, and display system. DAS hardware without software is useless and DAS hardware with poor software falls miserably short of the need for which the system is designed in the first place [21]. The software [21- 28] may consist of driver software and application software. The driver software is the layer of software that allows easy communication to the hardware. It works as an interface between application software and hardware. The driver software also prevents a programmer from having to do register level programming or complicated commands in order to access the hardware functions. The increasing sophistication of DAS hardware, computer and software continues to emphasize the importance and value of good driver software. The properly selected driver software can result in to good flexibility and performance with significantly reduced time required to develop the DAS application. The application software can be a development environment in which a custom application may be build.

Application software adds analysis and presentation capabilities to the driver software. Now a days, all the DAS hardware and driver software provides compatibility with commonly available development environments.

For the development of chemical gas lasers DAS, GeniDAQ software package may be selected to develop the application program. GeniDAQ is the most economical package from M/s Advantech Co. Ltd., which is best suited for Advantech's hardware. Moreover, the development of application is very easy as compared to other packages. This package is the best option when the sampling period requirement is of the order of few milliseconds. The other commonly available software packages are LabVIEW and DasyLab. A comparison between available software packages is given in Table (5.3).

Table 5.3. Comparison of software packages.

S. No.	Properties	LabVIEW	DasyLab	GeniDAQ
1	Company	National Instruments, Inc.	MeasX & Co.	Advantech Co. Ltd.
2	Latest Version	2012	Dasylab Release 12	4.11
3	Cost	Costliest	Costlier	Economical
4	Customization	Moderate	Moderate	Easy
5	Hardware	Yes	No	Yes
6	General	Modular	Package	Package

It is always better to write your own code in VC++ because in that case one can develop the tools and GUI the way one intends to perceive. In readymade packages, one has to use the general purpose tools/icons provided by the company. Fig. 5.6 and Fig. 5.7 show two GUIs developed in GeniDAQ and VC++ which clearly depict the difference between readymade and self developed GUIs.

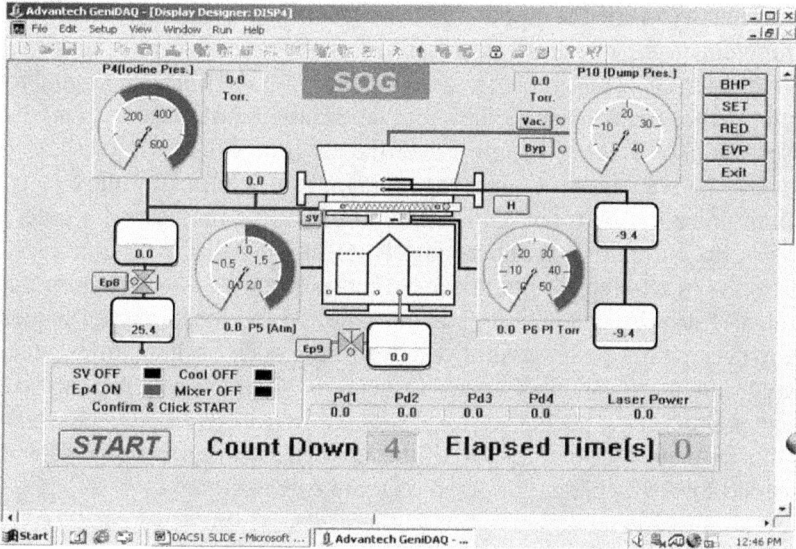

Fig. 5.6. Graphical user's interface of Singlet Oxygen Generator for COIL operation developed using GeniDAQ Software package.

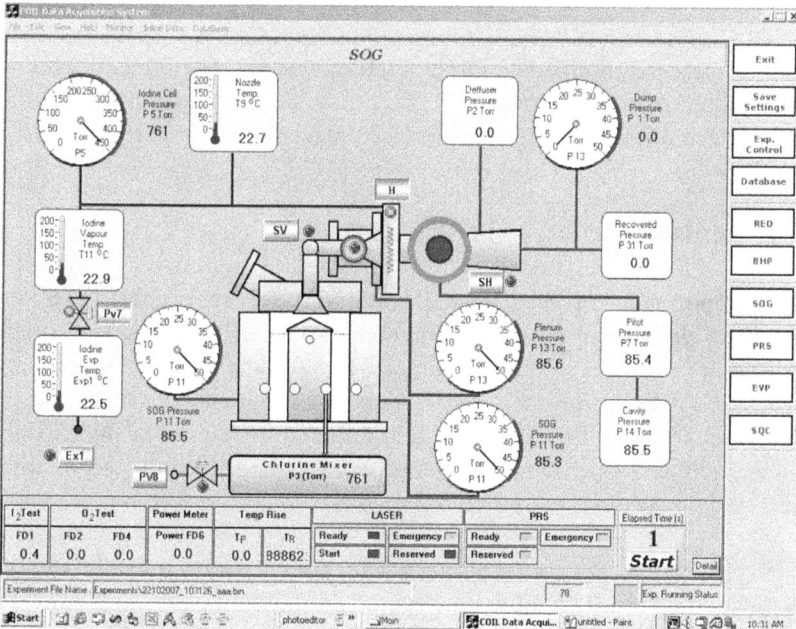

Fig. 5.7. Graphical user's interface of Singlet Oxygen Generator for COIL operation developed using VC++ Software package.

167

In order to make the application program user's friendly, design of graphical user interface is essential with several advantages and disadvantages. The graphical user interfaces (GUI) are important for computer systems because they provide symbolic interfaces, giving the user visual or graphical control over the program execution. GUI offers several advantages to an application program. It allows the communication between application program and the user in a natural symbolic form that is more closely resembles the human thinking process. Moreover, communication between application program and user is faster because the user is not required to enter program names. However, GUIs have several disadvantages such as it requires much larger memory and disk space to support graphic environment with higher processing speeds as large amount of data is required to move from hard disk to display. Also the amount of programming support needed for a graphic environment is more extensive.

Apart from the software package, the personal computer used in a data acquisition system can greatly affect the speed at which the data is continuously and accurately acquired, processed and stored for a particular application. When high-speed data acquisition is required, the selection of the processor speed and expansion bus (PCI or ISA) should be made accordingly to match the speed requirement of the application. The available memory should be large enough for acquired data storage requirements. The operating system is another important aspect that should be considered. This may be single tasking (e.g. MS-DOS) or multitasking (e.g. Windows 2000 or higher version).

5.4 Implementation of PC based DAS

The importance of a dedicated multipurpose real time DAS for chemical gas lasers and the requirements of DAS for a research grade gas lasers have been already discussed. A detailed system scheme required for DAS for operation and analysis of a chemical gas laser is shown in Fig. 5.8. The entire DAS structure may be divided into mainly three modules: operational and sequence control system, acquisition and analysis system, and safety interlocks.

The operational and sequential control system is responsible for the sequential and switching control of electrical/ electro-pneumatic valves/devices and online control and adjustment of physical parameters like flow rates of gases involved in lasing notion. This system also incorporates both analog outputs as well as digital output channels. The analog output channels are required for the flow rate control of different gas feed lines, whereas digital outputs are required for performing switching and sequential operations (on/off control) of various subsystems.

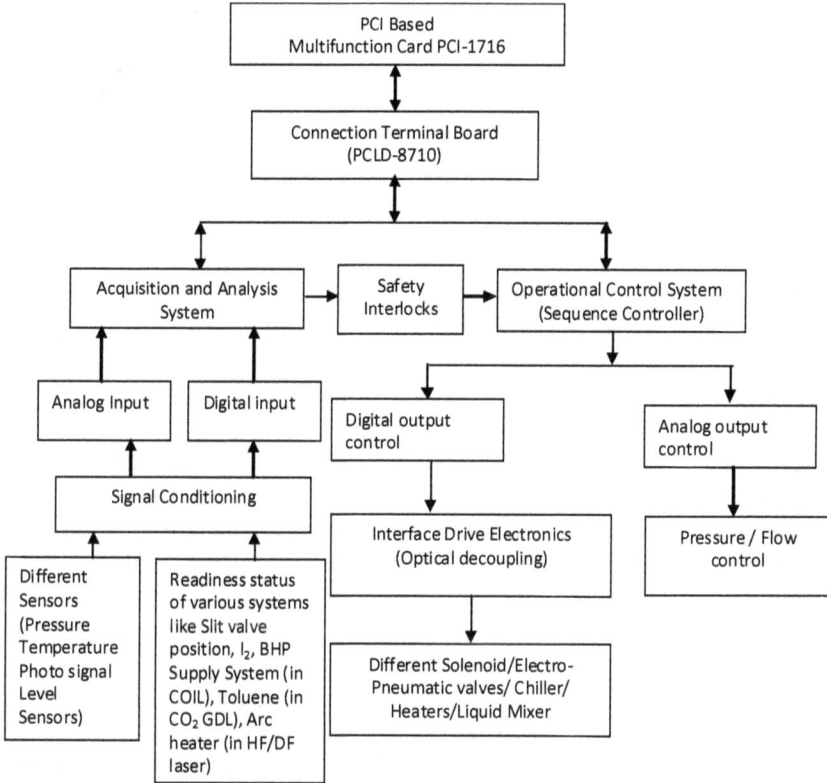

Fig. 5.8. Block diagram of DACS.

The acquisition and analysis system is concerned with the acquisition of analog and digital parameters along with the storage and online estimation of diagnostics parameters in graphical form for analysis of performance of different subsystems. The analog input channels are required for acquiring the signals from transducers like pressure sensor, temperature sensor and photodiode. On the other hand, the digital input channels are used to indicate the status of various subsystems such as the system readiness status of subsystems for laser firing and on/off status of various electro-pneumatic/solenoid /slit valves.

The safety interlocks is an essential part of chemical lasers for avoiding any accident and human hazards, which have been explained in detail in Chapter 3. This module utilizes input from different subsystems through the acquisition and analysis module and interacts to operational sequence control module for implementing necessary actions accordingly.

A detailed description of DAS hardware and its corresponding graphical user interface for a typical application in a COIL laser are taken up as an example in the following sections.

5.4.1 System Hardware

The basic building block of the DAS system is a multifunction data acquisition card. This data acquisition card interfaces the computer with the subsystems through custom-built electronics circuits. Advantech PCI-1716 data acquisition cards and Advantech–GeniDAQ software package/ application software developed in VC++ may be used to realize chemical gas lasers DAS [9-10]. In order to fulfill all the requirements for chemical gas lasers, as explained in earlier session, DAS may comprise of 150-200 channels depending upon the magnitude of parametric and control requirements. In one of our DAS, 170 channel DAS has been developed to operate COIL. It includes 64 analog inputs (AI), 10 analog output (AO), 32 digital input (DI) and 64 digital output (DO) channels. The development of the DAS system uses four PCI bus based PCI-1716 multi function cards from M/s Advantech, custom built interface electronic circuits like temperature to voltage converters, optically isolated digital output (DO) processing interface circuits, amplifier for photo-detectors etc. The channel requirements and their distribution according to their applications for complete operation of COIL are shown in Fig. 5.9.

PCI-1716 is a PCI bus based 16 bit high resolution multifunction card with 250 ksamples/s sampling rate. It includes complete functions for data acquisition and control including A/D conversion, D/A conversion, digital input, digital output and counter/timer. Followings are the specifications of this card:

- PCI-bus for data transfer;

- 16-channel single-ended or 8-channel differential A/D input channels;

- 16-bit A/D conversion resolution;
- 250 kHz sampling rate;
- Conversion time 2.5 µs;
- 2-channel D/A output;
- Driving capabiliy ±20 mA;
- 16-channel digital input;
- 16-channel digital output.

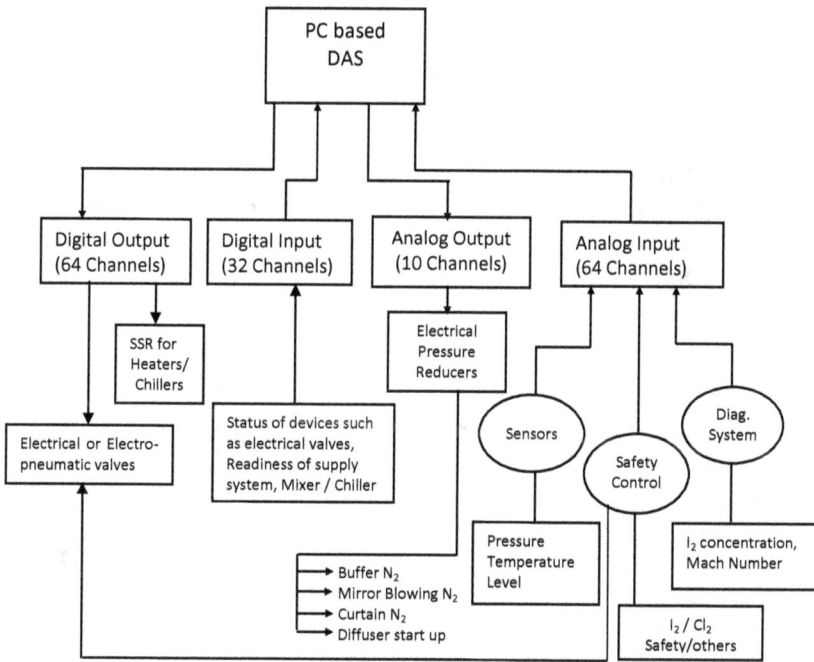

Fig. 5.9. Channel distribution of DAS.

The connections from the data processing card is an important consideration which should take care for convenient and reliable signal wiring. There are two possible options namely ADAM-3968 and PCLD-8710 for easy and reliable connections for PCI-1716 from Advantech. The PCLD-8710 Screw terminal board is employed for connections between data processing card and the equipments/subsystems because it provides additional features listed below as compared to ADAM-3968. The PCLD-8710 is a DIN-rail mounting screw-terminal board compatible with PCI-1716 card, which has 68 pin SCSI-II connector. It features the following functions:

- 2 additional 20-pin flat-cable connectors for digital input and output, which were utilized for connection between DO/DI channels and electrical devices to be controlled.

- Reserved space on the board to meet the needs for signal-conditioning circuits.

- Industrial-grade screw-clamp terminal blocks for heavy-duty and reliable connections.

The PCL-10168 shielded cable is used to interconnect the data processing card with the terminal boards. All the major electronics boards/ cards and other related devices are fitted inside a control panel which is shown in Fig. 5.10. The sockets are fitted on the bottom side of the panel from where the plugs along with the cable from various sensors and actuators are interfaced. The electrical connections among terminal boards, interface electronics modules and sockets are done inside the control panel.

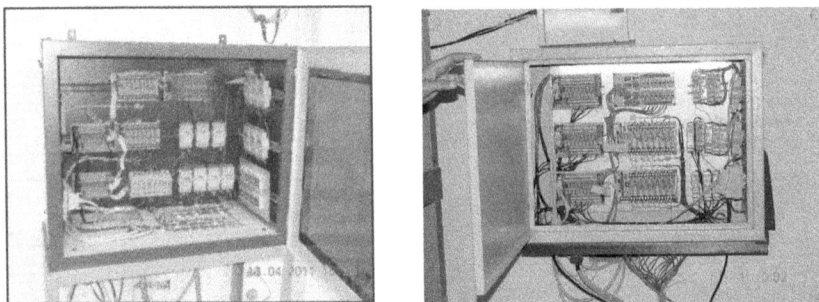

Fig. 5.10. Control panels of DAS designed for COIL gas laser.

5.4.2 Graphical User's Interfaces

Chemical gas lasers are large and complex system in which different subsystems are located at different positions. It is desired that a single person from control room should operate this laser. The number of displayed/monitored parameters at a time may be more than 100, which cannot be accommodated in a single window screen. Thus, management of the displayed parameters is essential. Moreover, it is required that the operator should feel as if he is in front of the system sitting inside the control room. Thus, to provide ease of operation, all the controls and crucial parameter display must be incorporated into five-six user's friendly bitmap control windows (graphical interface units) according to their operations.

This may be illustrated by taking an example of buffer/ diluent N_2 used for operation in all the three chemical gas lasers (COIL, CO_2 GDL and HF/DF lasers). It is used for several functions which are enlisted below:

1) Dilution of lasing and/or pumping species;

2) As a carrier gas;

3) Stabilization of pressure conditions in subsystems and laser cavity;

4) Mirror protection to prevent costly mirror optics;

5) Nitrogen curtain for catering boundary layer effects;

6) Diffuser start-up;

7) Pneumatic pressurization for electro-pneumatic devices etc.

These functions require its flow rate to be precisely regulated and on occasions online flow adjustment may also be needed. Hence, a dedicated graphical user interface is required to perform this function. Graphical user's interface must be designed as per the subsystem's operational, monitoring, effective display and control requirements.

Fig. 5.11 shows the GUI for N_2 buffer gas which has been implemented in case of COIL. As discussed in Chapter 2 under the section on Flow measurement technique, mass flow rate of gases can be controlled using orifice under choked flow condition. GUI allows the setting of electrical pressure reducer and displays the line pressures of the important lines in addition to the operation of corresponding electrical valves. If upstream pressure of orifice can be controlled, flow rate of gas can be controlled. GUI of Fig. 5.11 shows several gas feed lines in which each line comprises of analog output voltage adjustment tool. The variation in analog voltage to electrical pressure reducer can be adjusted by scroll bar and corresponding pressure is displayed which is acquired by analog input channel of DAS. The display of pressures of gas feed line is displayed by dial gauges as well as in digital form. The timing control is implemented by control buttons which can be used to open electrical valve by pressing it or it can take pre defined time from sequence setting GUI. These control buttons are linked with digital output of DAS followed by DO processing interface electronic circuit to meet current/voltage requirements of electrical valves. The on/off status of electrical/electro-pneumatic valves is indicated by green/red indication on screen which is linked to digital input of DAS. A similar GUI may be designed for other chemical lasers as well.

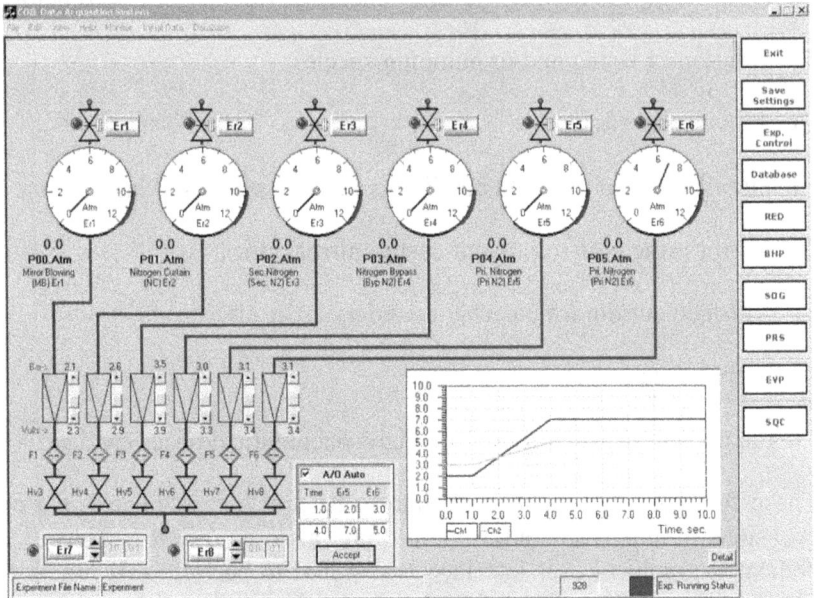

Fig. 5.11. Graphical user interface for gas flow control in chemical gas lasers.

Similarly, all the gas lasers require GUI for timing sequence control. The setting of on/off timing of different events (supply of different chemicals/gases and switching on/off of different solenoid/electro-pneumatic valves and other electrical subsystems demand a GUI in which operator can feed the on/off timings of various events during laser operation. Fig. 5.12 shows one of sequential control GUI which may be implemented in sequential control of these lasers.

Similarly depending upon gas lasers, one can design other GUIs like in COIL; other GUI can be for iodine supply system, BHP supply system etc. All the GUIs should have toggling switch to switch over from one window to other windows as visible in above examples shown in Fig. 5.11 and Fig. 5.12.

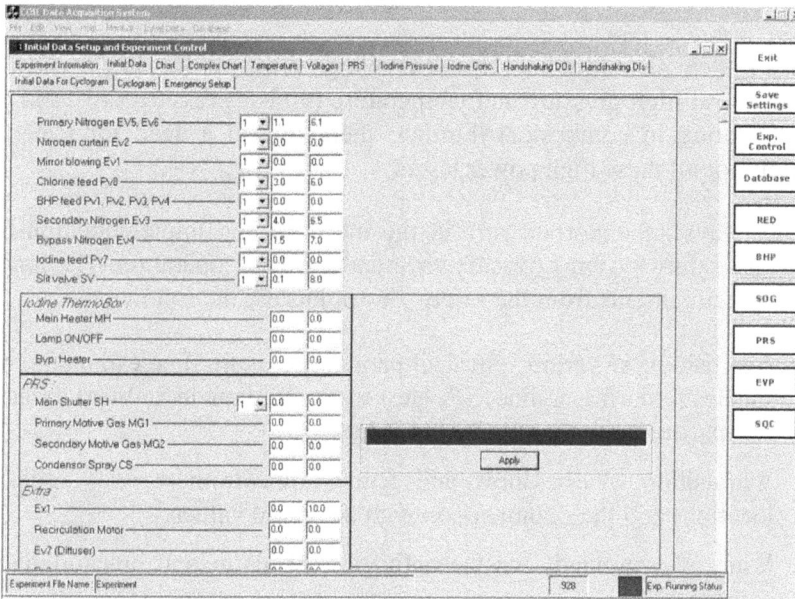

Fig. 5.12. Graphical user interface for sequential in chemical gas lasers.

5.4.3 Safety Interlocks

Most of the gas lasers utilize various hazardous gas effluents and chemicals. In order to exploit full potential of these lasers, one must take diligent care of the safety issues associated with the handling of these chemicals and the involved processes. Therefore, there is a need to employ suitable detection systems for these gases and chemicals along with safety schemes for safe operation [29] of these lasers. Further, it is essential that the manner of implementation of various checks on all the safety concerns must be addressed at the time of laser development itself. In case of COIL, Cl_2, I_2 and BHP are of great concern as far as safety is concerned. However, this being a low pressure and low temperature system, it does not require high pressure safety. Similarly, HF/DF requires SF_6 safety however it is not as hazardous as free F_2. F_2 is produced on line by SF_6 so safety requirement are less stringent as compared when F_2 is being handled directly. In HF/ DF laser hydrogen safety also needs to be taken in to account. Hydrogen gas is colorless, odorless but inflammable. It forms flammable and explosive mixtures with air over a wide range of concentrations.

This system also does not require very high pressure. But in CO_2 GDL, operates at high pressure and temperature (40 bar pressure and 1800 K temperature in combustor). Further, there is also a need for general safety for all these high power lasers.

DAS plays an important role in the interface and implementation of various safety systems for safe chemical gas laser operation. The main DAS requirements from the safety viewpoints are the following:

- Acquisition of various physical parameters and online estimation of species (chlorine, iodine, SF_6 etc.) concentrations in ambient air and comparison with their threshold values.

- Acquisition of pressures and temperature from various critical locations and their comparison with threshold values.

- Issue of commands for the activation of alarm and precautionary equipments.

- Determination and display of various relevant parameters based on the pre-defined mathematical formulations.

In order to explain the safety of gas lasers, let us consider the case of COIL. In COIL operation, one has to use chlorine gas and iodine gas, which are highly toxic along with hydrogen peroxide solution, which is highly explosive in pure form at high concentration and extremely hazardous for human being in case of contact. Hence, the major safety considerations in COIL are that of chlorine and iodine leakage, handling and storage of hydrogen peroxide solution. In addition, excessive pressure development in the subsystems like singlet oxygen generator, iodine chamber etc. are also areas where safety is a concern and safety aspects are required to be critically examined as these are potential points of leakage of these gases, which may also affect the laser performance.

Chlorine gas is highly toxic in nature, which adversely affects human respiratory system. It irritates the mucous membrane, respiratory tract and eyes. Chlorine produces pallor cold, clammy skin and weak pulses. It also results in difficulty in breathing.

The criticality is the combined effect of the concentration and exposure time. Maximum allowable limit for prolonged exposure is 1 ppm whereas 4 ppm is the maximum allowable limit for ½ to 1 hour. For throat irritation and coughing, the least amount required is 15.1 ppm and 30.2 ppm respectively.

Iodine vapors are intensively irritating to mucous membranes and adversely affect the upper and lower respiratory system [30]. Inhalation of iodine vapor leads to excessive flow of tears, tightness in the chest, sore throat and headache. It increases pulmonary flow resistance, decreases compliance and the rate of ventilation [31]. Iodine vapors are highly toxic and humans can work undisturbed at 0.1 ppm; with difficulty at 0.15–0.2 ppm and it is impossible to work at concentrations of 0.3 ppm. The odor threshold has been reported at 0.9 ppm, so irritation may occur before the odor is detected. The permissible exposure limit (PEL) is 0.1 ppm (as a ceiling value). Most of the countries (USA, Australia, Germany, etc.) have the same exposure value. The immediately dangerous to life and health (IDLH) value is 2 ppm.

In order to handle any kind of leakage situations in COIL operation, appropriate safety measures need to be implemented by employing suitable sensors. The sensors are required to be interfaced with suitable alarming system and provide commands for auto switch on of the necessary protection equipment such as air ventilation system. In order to ensure safe operation of the COIL system, three major safety systems that must be implemented are:

1) Chlorine leak detection and air ventilation system;

2) Iodine leak detection system and air ventilation system;

3) BHP safety and monitoring of other critical parameters and their safety issues (such as: excessive pressure development at critical locations).

Fig. 5.13 shows the general scheme that may be followed by all the safety interlocks. The diagram shows that the sensor signal obtained from all the critical systems is continuously compared with set reference signal and in case of sensor signal exceeding the set threshold value, the safety system is operated. The implementation of these systems with DAS is discussed in details in the following sessions.

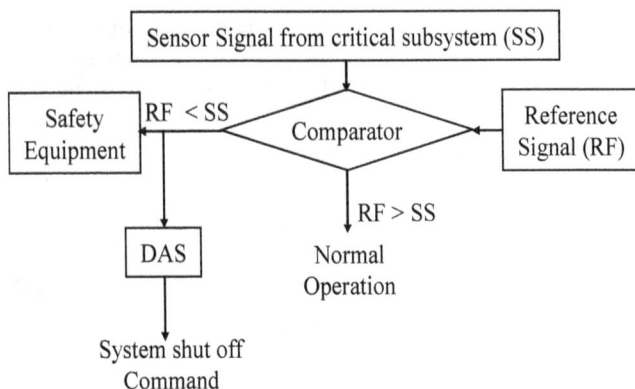

Fig. 5.13. General scheme for Safety Interlocks for chemical gas lasers.

5.4.3.1 Chlorine Safety Measures

One of the basic safety precaution that may be incorporated into the system is that the operating pressure of the Cl_2 feed lines should be maintained at a pressure less than the atmospheric pressure, typically 550 –600 torr. This prevents outflow of the chlorine to the ambient in case of leak points existing in the system. It will be the ambient air that enters into the feed lines rather than Cl_2 leaking out.

For various reasons, in laboratory COIL systems, the chlorine supply system is kept in an isolated room and only the feed line is taken to the main laser room. Both of these rooms need to be equipped with the detection system and suitable precautionary air ventilation system. The chlorine leak detection system (sensor and electronics together) should have a fast response time and high sensitivity to cater to the emergency situation efficiently. An electrochemical sensor (ADVANCE 200), which uses special materials of construction suitable for chlorine environment for improved lifetime and reliability, may be used. The sensor produces a current output signal according to the concentration it observes and the sensor unit has an option to set a threshold concentration of 1 and 10 ppm and produce a 5 mA current output for any concentration above the set threshold value. This sensor provides signal to the control unit to activate air circulation system when Cl_2 concentration exceeds the threshold level. Air circulation system capacity is typically designed in such a way so as to cater to the ratio of nearly 10:1 of room capacity.

The ventilator output is required to be passed through a scrubber (charcoal) to absorb the chlorine molecules before it is exhausted out of the chimney into the atmosphere. The entire operation time from the time the leak is detected to switching on the air ventilation system is required to be as short as possible, typically less than 1 sec.

5.4.3.2 Iodine Safety Measures

The basic safety precautions that may be incorporated into the system is that the operating pressure of the I_2 feed lines should be maintained at a pressure less than the atmospheric pressure, typically 100-150 torr. This prevents outflow of the iodine to the ambient in case of leak points existing in the system. It will be the ambient air that enters into the feed lines rather than I_2 leaking out.

One can use electrochemical sensor (TA-2102 Iodine I2 Gas Detector) for iodine detection. This sensor has detection range from 0 to 5 ppm with 0.1ppm resolution. This sensor provides signal to automatic control unit and DAS to activate air circulation system and alarm when I_2 concentration exceeds the threshold level. Its transmitter operates at 24 V dc and produces 4-20 mA current output which is linear. This current output is fed to a current to voltage converter to get a voltage output signal and is compared with a reference signal to activate air circulation system and alarm similar to chlorine. The ventilator output can be passed through a liquid nitrogen trap to adsorb iodine molecules before the air is exhausted out of the chimney into the atmosphere.

The electronic interface and precautionary air ventilation system design are the same as that for chlorine. However, in case of iodine air circulation capacity may be lower and is normally taken as 5:1 of the room capacity.

Due to the highly irritating nature of iodine, respiratory protection must be used whenever the PEL will be exceeded. Since odor detection is unlikely below the PEL, respiratory protection has been used whenever handling iodine at a clandestine lab. Air purifying respirator (APR) cartridges are certified by National Institute for Occupational Safety and Health(NIOSH) to protect the worker up to a maximum limit, defined as the maximum use concentration (MUC).

5.4.3.3 Safety for Hydrogen Peroxide, BHP and Toluene

Hydrogen peroxide and BHP are used in COIL as a liquid fuel whereas Toluene is used as a liquid fuel in CO_2 GDL system. Normally, the hazard due to hydrogen peroxide increases with increase in its concentration. As the concentration increases the explosive nature increases and makes its storage and transportation extremely complicated. Hence, the concentration of commercially available hydrogen peroxide solution is 30-50 % with the rest being water. Inhalation of H_2O_2 vapors or mist is irritating to the respiratory tract. The concentration of 5 % w/w and above can cause irritation or burns, with severity increasing with concentration. With solution of 6 % w/w and above can damage eyes permanently. National Institute for Occupational Safety & Health (NIOSH) [32] permissible exposure limit (PEL) is 1 ppm (1.4 mg/m^3 air) for a normal 8 hour working day. The decomposition of H_2O_2 occurs with increase in temperature. During decomposition, H_2O_2 continuously decompose in to water and oxygen, high gas pressure may develop in sealed container. Hence, it must be stored suitably vented containers and never in hermetically sealed containers. The temperature of storage tank should be kept under monitoring. Only trained personnel with protective equipments (protective clothing, gloves and mask) should be allowed to handle H_2O_2.

In order to achieve optimal COIL performance, correct BHP composition is critical typically being KOH: H_2O_2 molar ratio: 6.5:7. At the same time the water content in H_2O_2 also decides the extent of water vapor generation during the reaction. In order to minimize the water vapor production the composition is optimized and the solution is used at a possible low temperature without freezing. We have used 50 % H_2O_2 in our COIL device and the BHP solution is maintained at a temperature of about -20 °C. BHP tends to have an explosive nature due to cascading thermal runaway as the temperatures rise above -6 °C. Hence, a constant temperature monitoring system interfaced with PC needs to be provided as a safety interlock. Whenever, temperature rises beyond -10 °C the sensor unit produces a command for alarm and shut down of the laser operation. The chiller system for the cooling of BHP is also simultaneously activated and in case the BHP temperature is still not maintained, the command for switching on the solenoid valve for the drain line is issued for BHP disposal.

Toluene is used in CO_2 GDL as a liquid fuel because it is much less toxic as compared to benzene. Toluene should not be inhaled due to its health effects. Inhalation of toluene vapor can affect the central nervous system (CNS). At approximately 50 ppm, slight drowsiness and headache have been reported. Irritation of the nose, throat and respiratory tract has occurred between 50 and 100 ppm. Concentrations of about 100 ppm have caused fatigue and dizziness; over 200 ppm has caused symptoms similar to drunkenness, numbness, and mild nausea; and over 500 ppm has caused mental confusion and incoordination. Higher concentrations (estimated at higher than 10000 ppm) can result in unconsciousness and death. Most serious incidences of exposure have occurred when the vapor has accumulated in confined spaces.

Precautions should be taken while handling toluene. It should be kept away from heat as well as from sources of ignition. One should not breathe gas/fumes/ vapor/spray of toluene. Protective clothing should be worn. In case of insufficient ventilation, one should wear suitable respiratory equipment. If ingested, seek medical advice immediately and show the container or the label. Avoid contact with skin and eyes. Keep away from incompatibles such as oxidizing agents. It should be stored in a segregated and approved area. Keep container in a cool, well-ventilated area. Keep container tightly closed and sealed until ready for use.

5.4.3.4 Safety for HF/DF Laser

HF/DF laser involves reaction of fluorine atoms with hydrogen or deuterium to generate vibrationally excited HF or DF molecules, which are utilized for generation of laser energy in the cavity. Fluorine is a toxic and hazardous gas and involves specialized handling. However, in arc-driven HF/DF laser, generation of fluorine atoms is done in-situ by thermal decomposition of sulphur hexafluoride. Therefore, storage and handling of free fluorine may be avoided. Gas detection unit should be used for detecting gas leakage. It should be interfaced with a safety system to issue command using operational control module to switch on alarm as well as interrupt operation of laser in case of leakage. The general scheme of safety interlocks has already been shown in Fig. 5.13. The detection of fluorine gas may be carried out by using Universal toxic gas transmitter (4600 series) which can detect up to 10 ppm level. Permissible Exposure limit (PEL) for fluorine is 0.1 ppm as per Occupational Safety & Health Administration (OSHA). However, detection limit should be set at the threshold of 0.05 ppm. A similar air ventilation scheme with trap may be used in case of fluorine leakage as in iodine/chlorine leakage case (explained in earlier sections). The detection of hydrogen leak may be carried out by using model number TGS 821. This sensor has high sensitivity to hydrogen gas and can detect concentrations as low as 50 ppm. In case of HF, OSHA PEL is 3 ppm and Polytron L 20 HF gas sensor (2180 dragger) may be used for its detection and accordingly threshold may be set in safety interlocks. SF6 is relatively non toxic gas used in a number of applications (circuit breakers etc) due to its qualities. The OSHA PEL is 1000ppm and a number of SF_6 gas sensors (Model number GS860ASW) are available commercially.

5.4.3.5 Safety for High Pressure in CO_2 GDL

CO_2 GDL system operates at high pressure (35 bar) and high temperature (1700 K). In order to operate this laser, air gas supply pressure of ~ 180 bars is required; hence, it needs safety attention mainly because of this high pressure. The same safety scheme (as low pressure safety implemented in COIL) may have been implemented for catering the high pressure emergency situation by using safety relief valves and sound alarm signal. In addition, all gas supply cylinders and valves should be securely shutoff when not in use and pipelines should be periodically checked for any possible leakages. Safety relief valves in the high pressure system must be installed at various important locations and they must be regularly checked for proper functioning. Only trained and experience personnel should be allowed to handle the high pressure system.

5.4.3.6 Safety in Laser Operation

All the three gas lasers are complex system with the various sub systems being spread over a wide area. Typically, the prime parameters which govern the laser performance to a large extent are the optimal pressure and temperature conditions which mainly control the flow morphology in this flowing medium gas laser. Since it involves use of various potentially hazardous chemicals, it is not only essential to continuously monitor the pressure and temperature at all the critical locations for optimizing laser performance but also from a safety stand point.

In the laser head, there is possibility of pressure development at various critical locations such as: singlet oxygen generator (where Cl_2 is injected), buffer gas supply system, iodine chamber and iodine injection cell. The pressure at all the relevant locations is monitored continuously using diaphragm type sensor suitable for a corrosive working environment. The maximum acceptable limits of pressure at each individual location are pre-defined in the PC and increase beyond these stated limits activate the safety interlocks for necessary action, which are discussed in later sessions. These actions are essential to ensure personnel and equipment safety and extreme care needs to be taken in prescribing the acceptable limits and implementation of safety action.

The height of laser beam is kept above the height of a normal human being and the laser is operated remotely with no personnel being allowed inside the tunnel room during laser operation. All the three chemical gas lasers radiates in infrared region (1.315 μm in COIL, 2.7-3.4 μm in HF/DF laser and 10.6 μm in CO_2 GDL). The laser radiations at these wavelengths can cause both eye as well as skin damage, but eye damage is more serious. For low power lasers at this wavelength, eyes can be guarded using laser goggles but in case of high power lasers of Class IV grade that have been discussed here, these are of not much use. Even the scattered radiations from the walls etc. are sufficient to cause damage. So even during development, one has to be very careful. We have tested the operation of laser by housing the laser head in a separate closed room fitted with CCD cameras, from where all the activities are monitored remotely. The laser beam has been either dumped into cone shaped calorimeter for laser power measurement or allowed deliberately to fall on various targets to study the effect of laser beam. Both the calorimeters as well as the targets were well shielded with thick blackened metal sheets from all sides. Further, audio and visual alarm systems have also been implemented during the testing of lasers from safety point of view.

5.5 Performance Testing of Chemical Gas Lasers

All the sensors, measurement techniques, interface circuits and data acquisition system so far discussed can appropriately be utilized to test the performance of chemical gas laser systems and to optimize them as well. In this section utility and importance of measurement techniques in development of gas lasers has been elaborated. All the gas lasers can be performance tested with the help of three kinds of experiments:

1) **Cold runs:** In cold run experiments, the actual fuel (lasing as well as pumping medium) is replaced with equivalent buffer gas and pressure and flow rate conditions are optimized in different subsystems of laser. For example in case of CO_2 GDL, combustor pressure and flow conditions are optimized with the help of air only. No toluene and gasoline is used for initial optimization. Similarly, in HF/DF laser, cavity and plenum conditions are achieved with the help of buffer nitrogen gas. No SF_6 or H_2 is used. In the same way, in COIL, BHP solution is replaced with water and nitrogen gas is used to compensate the entire gas flow (chlorine, buffer and iodine). Thus, water and nitrogen are passed through the system in order to optimize the flow parameters and to verify the flow conditions expected in individual subsystems. The SOG operational pressure, gas velocity, plenum conditions, cavity Mach number are to be determined for various flow conditions through these experiments. The increase in dump pressure is also essential to observe in order to optimize the run time of the system.

2) **Hot runs:** In hot run experiments of COIL, liquid BHP solution and chlorine is passed through the system and singlet oxygen measurement is done. Nitrogen and iodine is allowed to pass through the corresponding injection locations, mainly aiming to optimize the iodine concentration in these experiments. Here the secondary main and secondary bypass nitrogen gases are adjusted between runs to regulate the iodine flow into the laser system. Iodine concentration may also fine tuned by adjusting the iodine evaporator temperature. In case of CO_2 GDL, toluene and gasoline is used in combustor firing to optimize pressure (~40 bar) and temperature (1700-1800 K) in the combustor chamber and achieve the cavity conditions 30-35 torr via adiabatic expansion of gases through supersonic nozzle. Similarly in HF/DF laser, all the fuel feeds are given and optimized. In all the three lasers, in the hot runs small signal gain measurement may also to be done in the laser cavity without extracting laser power i.e. resonator is not installed.

3) **Power runs:** These runs are actual power runs in which laser power extraction is carried out. During these runs, in COIL, the actual fuels like liquid BHP, chlorine, iodine and buffer gases are injected as per the optimizations carried out in cold and hot runs. The major aim of these experiments is to optimize the flow rates to produce maximum power output.

In order to present to the reader a glimpse of the manner in which DAS may be utilized as a tool to analyze system performance, illustrative examples from our experience in terms of experiments conducted on COIL system are discussed briefly.

5.5.1 Direct Parameter Analysis

The basic parameter that serves as a window into a COIL system is the stable pressure achieved in the various critical subsystems of the laser module. In case of COIL few important ones are: SOG, cavity, Pitot pressure and supply pressures for diluent nitrogen and chlorine. A temporal plot of SOG, cavity and Pitot pressures for a typical COIL run is shown in Fig. 5.14. The timings and qualitative variation of supply diluent nitrogen and chlorine pressures are superimposed on the same plot. Since the supply pressures are of considerably higher values than the pressures inside the COIL tunnel. The typical value for diluent N_2 is 4.5 bar and for Cl_2 is 600 torr.

One of the most noticeable things evident from DAS display for temporal graph shown in Fig. 5.14 is that stabilization time of generator pressure is nearly 5.5 s. This is substantially a long time in the context of a high power laser system. It would be of distinct benefit if the stabilization time may be reduced by employing some kind of alteration in the sequence. Hence, a ramp variation in the supply of primary N_2 was tried so as to be able to stabilize the SOG pressure faster (as shown in Fig. 5.15).

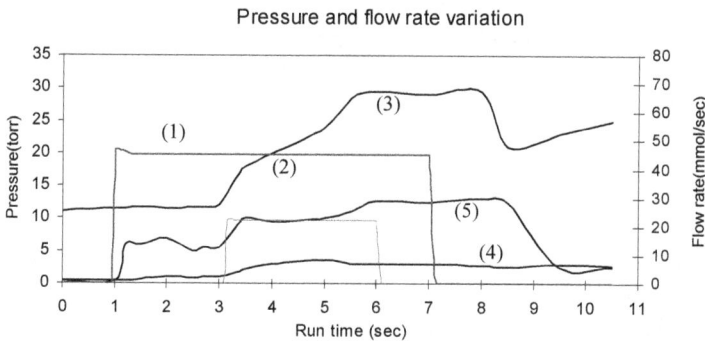

Fig. 5.14. The temporal variation of diluent nitrogen (curve 1) and chlorine (curve 2) flow rates along with generator, cavity and Pitot tube pressures (curves 3,4 and 5 respectively), red is chlorine pressure during COIL run without diluent nitrogen ramp.

Pressure and flow rate variation

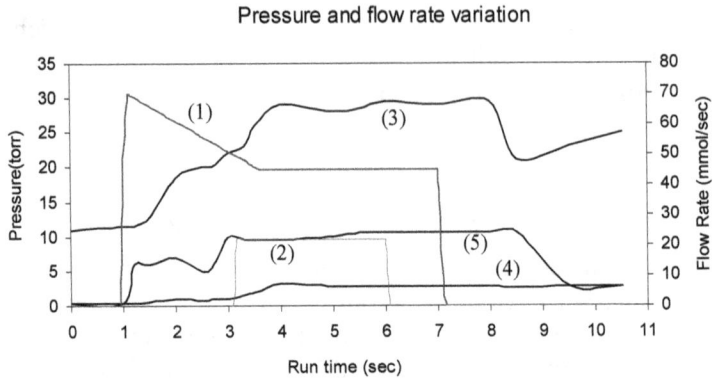

Fig. 5.15. The temporal variation of diluent nitrogen (curve 1) and chlorine (curve 2) flow rates along with generator, cavity and Pitot tube pressures (curves 3,4 and 5 respectively) during a COIL with diluent N_2 ramp

It is apparent from the two plots that the SOG pressure stabilizes faster. The stable duration operation for SOG in the first case without ramp(Fig. 5.14) is nearly 2.5 s, whereas in the later case (Fig. 5.15), it is approximately 4.5 s. Hence, the provision of ramp is of great help for better laser performance, and an online gas flow variation can only be executed using a DAS.

5.5.2 Derived Parameter Analysis

A couple of derived parameters, viz., iodine flow rate and Mach number and their influences on laser operation are discussed in the present session. Both of these parameters are important as the former is the basic lasing species and the latter enables production of necessary conditions for lasing action inside cavity.

5.5.2.1 Iodine Flow Rates

The measurement diagnostics for iodine flows has already been discussed in Chapter 3, which uses optical absorption method at 490 nm. DAS in real time acquires the temporal variation of absorption signal corresponding to concentration and iodine cell pressure. The iodine partial pressure and iodine flow rate are subsequently estimated and displayed on the monitor in real time. The data acquisition card acquires the corresponding data from photo detector, pressure and temperature sensors via different analog input channels of data acquisition card and stored in the different files in the software, which are used for the online estimation of the iodine flow rate. Fig. 5.16 shows typical temporal curves of iodine signal of optical cell. The temporal variation of iodine partial pressure and iodine flow rates are shown in Figs. 5.17 and 5.18 respectively.

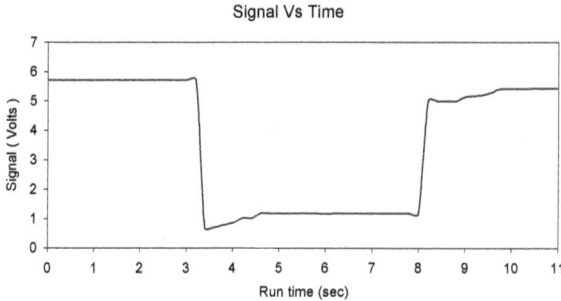

Fig. 5.16. Temporal variation of iodine photodiode signal in COIL runs.

The above curves are for typical COIL operation in a 500 W class COIL system [33]. The required iodine flow rate is about 2 % of the chlorine flow rate (27mmol/s). Hence, the steady iodine flow rate for optimal conditions is required to be nearly 0.4 –0.5 mmol/s. The above show optimized run parameters for iodine system.

However, it would be correct to mention that several experiments were performed to reach these parameters. Since, in many of our runs we experienced decay in iodine flow rates with time, which is undesirable from point of view of laser operation. The plot is shown in Fig. 5.19 for better comprehension.

The observation of iodine flow decay in Fig. 5.19 was attributed to reason that the carrier gas supplied into the evaporator flushes out maximum iodine vapors during the initial duration since they are at saturated conditions inside the volume, and thereafter the iodine available is because of the surface evaporation. Hence, in order to address this issue high power tungsten lamp setup is implemented on the top of the evaporator. The whole idea is to switch on the lamps so as to improve the surface evaporation rate to prevent fall in iodine concentration. The effect of its use is easily visible in Fig. 5.18. Thus, DAS again exhibits is usefulness from viewpoint of laser optimization.

Iodine pressure Vs Time

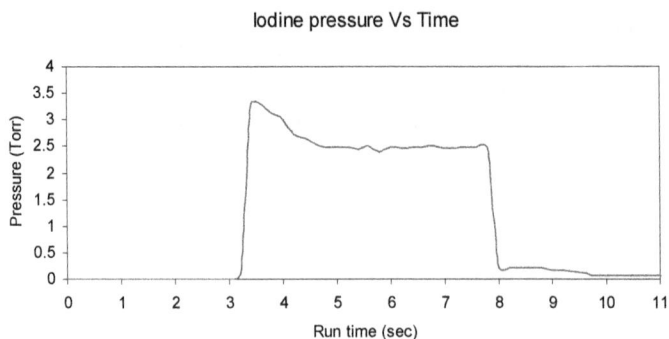

Fig. 5.17. Temporal variation of iodine partial pressure in COIL runs.

Iodine flow rate Vs Time

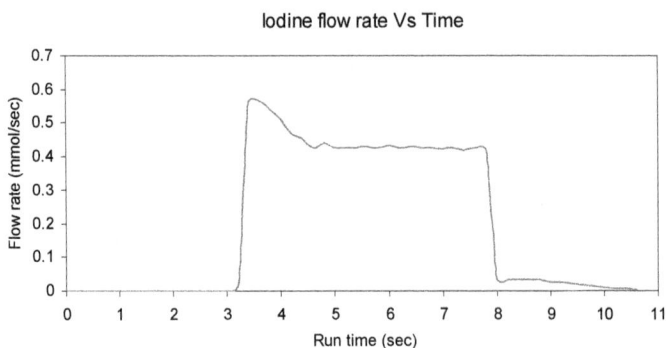

Fig. 5.18. Temporal variation of iodine flow during COIL power
run with lamp.

Iodine flow rate Vs Time

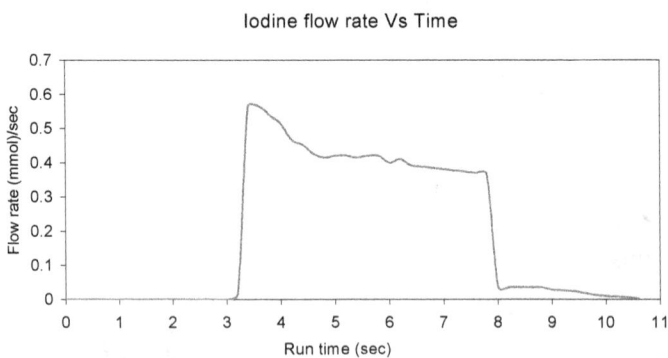

Fig. 5.19. Iodine flow rate decrease without lamp during COIL power run.

188

5.5.2.2 Mach Number

Another example of derived parameter variation is Mach number. Various methods for determination of cavity Mach number have been detailed in Chapter 3. Mach number supposedly has direct impact on the extracted power in case of COIL. A supersonic Mach number is essential at all times inside the cavity when power extraction is being carried out.

In order to illustrate the effect of Mach number on COIL power[34] two plots Fig. 5.20 and Fig. 5.21 are shown. The former shows a stable Mach number of close to 1.5 in the entire run duration leading to stable power output. However, in the second case the cavity Mach number drops to below sonic leading no power condition after 4 sec of run time. The strong impact of Mach number theoretically expected on COIL power is hence confirmed by the experimental observations.

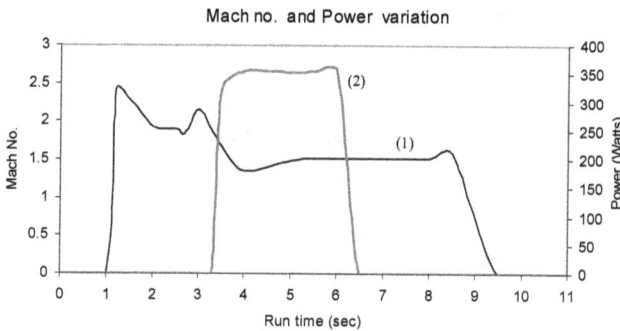

Fig 5.20. Temporal variation of Mach number (curve 1) and COIL output power (curve 2) for a typical successful COIL run.

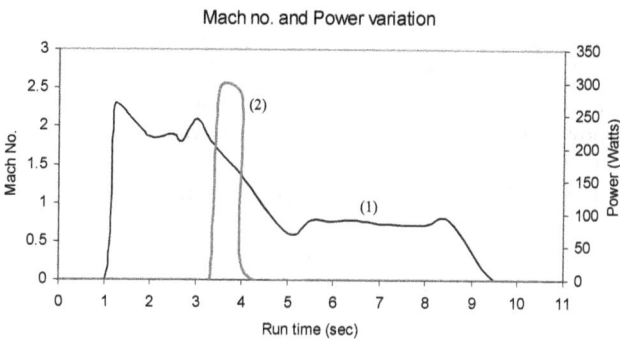

Fig. 5.21. Temporal variation of Mach number (curve 1) and COIL output power (curve 2) for an unsuccessful COIL run.

189

In closure it would be appropriate to state that similar analysis and suitable inferences may be drawn in case of HF/DF and GDL laser systems by designing an effective DAS.

References

[1] James W. Dally, Rilley W. R. and McConnell K. G., Instrumentation for Engineering Measurements, Chapter 4, *John Willy & Sons, Inc.,* 1993.

[2] http://www.plc.com

[3] http://www.industrialtext.com

[4] Ian. G. Warnock, Programmable controllers-operation and application, *Prentice-Hall,* 1988.

[5] Murphy N. E., General purpose data acquisition and control via the IBM PC Centronics printer port, *Meas. Sci. Technol.,* 7, 1996, p. 203.

[6] Strether S., Commercial-Off-The-Shelf (COTS) Software Systems for Data Acquisition and Analysis, *Sound and Vibration,* 1996, p. 2.

[7] Gerd Wostenkuhler, Data Acquisition Systems (DAS) in General, Handbook of measuring system design, *John Wiley & Sons, Ltd,* 2005.

[8] Howard Austerlitz, Data acquisition techniques using PCs, 2nd ed., Academic Press, *Elsevier Science,* USA, 2003.

[9] Mainuddin, R. K. Tyagi, R. Rajesh, Gaurav Singhal and A. L. Dawar, Real-time data acquisition and control system for a chemical oxygen-iodine laser, *Measurement Science and Technology,* 14, 2003, p 1364.

[10] Mainuddin, Gaurav Singhal, R K Tyagi, Data acquisition system for flowing gas lasers, V2-184-188, in *Proceedings of the 3rd IEEE International Conference on Electronics Computer Technology (ICECT' 2011),* Kanyakumari, 2011.

[11] Mainuddin, Gaurav Singhal, R. K. Tyagi, and A. K. Maini, Diagnostics and data acquisition for chemical oxygen iodine laser, *IEEE Transactions on Instrumentation and Measurement,* 61, 6, 2012, p 1747.

[12] http://www.modicon.com

[13] http://www.aut.sea.siemens.com

[14] http://www.standardautomation.aa.psiweb.com/specs/cutler.shtml

[15] Mainuddin, M. T. Beg, Moinuddin, R. K. Tyagi, R. Rajesh, Gaurav Singhal and A. L. Dawar, Real time gas flow control and analysis for high power infrared gas lasers, *International Journal of Infrared and Millimeter Waves,* Vol. 1, Jan. 2005, p. 91.

[16] Hoeschele, D. F., Analog-to-Digital/Digital-to-Analog Conversion Techniques, *John Wiley Interscience,* 2nd ed., 1994.

[17] Spencer, C. D., Digital Design for Computer Data Acquisition, *Cambridge Univ. Press,* London/New York, 1990.

[18] Park J. and MacKay S., Practical Data Acquisition for Instrumentation and Control, *Newnes,* Oxford, 2003.

[19] Sheer D., Monolithic high resolution ADCs, *EDN* May 12, 1900, p. 110.

[20] O'Leary S., Self calibrating ADCs offer accuracy, flexibility, *EDN*, June 22, 1995, pp. 77-85.
[21] http://www.ni.com
[22] http://www.tmagilent.com
[23] http://www.datatranslation.com
[24] http://www.dataq.com
[25] http://www.keithley.com
[26] http://www.labtech.com
[27] http://www.dasytec.com
[28] http://www.advantech.com
[29] Mainuddin, Gaurav Singhal, RK Tyagi, AK Maini, Development of safe infrared gas lasers, *Journal of Optics and Laser Technology*, Vol. 43, 2013, p. 56.
[30] Documentation of the Threshold Limit Values and Biological Exposure Indices, 6th Edition, American Conference of Governmental Industrial Hygienists, *Cincinnati, Ohio,* 1991, pp. 799-800.
[31] Cassarett, L. J., Toxicology of the Respiratory System. In: Toxicology, The Basic Science of Poisons, *Macmillan*, New York, 1975, pp. 201-224.
[32] NIOSH, NIOSH pocket guide to chemical hazards, Cincinnati OH: US, Department of health & Human Services, Public Health Service, Centre of deases control, *National Institute for Occupational Safety & Health*, DHHS (NIOSH), Publication No. 94-116, 1994.
[33] Mainuddin, M. T. Beg, Moinuddin, R. K. Tyagi, R. Rajesh, Gaurav Singhal and A. L. Dawar, Optical spectroscopic based In-line iodine flow measurement system-an application to COIL, *Sensors and Actuators*: B, Vol. 109, 2005, 2005, p. 375.
[34] Mainuddin, M. T. Beg, Moinuddin, R. K. Tyagi, R. Rajesh, Gaurav Singhal and A. L. Dawar, Real time gas flow control and analysis for high power infrared gas lasers, *International Journal of Infrared and Millimeter Waves*, Vol. No. 1, Jan. 2005, p. 91.

Chapter 6

Analysis of Uncertainty in Measurement

A measurement is a process whereby the value of a physical variable or parameter is estimated. All measurements are accompanied with error [1]. The lack of knowledge about the sign and magnitude of measurement error is termed as measurement uncertainty. A measurement uncertainty estimate is the characterization of what we know statistically about the measurement error. Therefore, a measurement result is only complete when accompanied by a statement on the uncertainty in that result. Hence, in case of chemical lasers as well it is essential to characterize the data collected for the host of parameters in terms of the associated uncertainties with them.

The typical accepted procedure [2] for estimating uncertainty and applicable for the case of chemical lasers involves the following steps:

a) Define measurement process;

b) Identify sources of error;

c) Generate model;

d) Estimate uncertainties.

The present chapter discusses these aspects with reference to measurements carried out in case of chemical lasers and few typical examples for estimation of uncertainty are presented towards the end.

6.1 Uncertainty Methodology

The first three points in uncertainty analysis viz., definition of measurement process, generating an error model and sources of error are included in this session as they are closely related to one another.

6.1.1 Define Measurement Process

The foremost step to uncertainty analysis is to categorize the physical parameter measured. The parameter may be a directly measured value or derived from measurement of several other physical variables. The definition that we had outlined in Chapter 2 for "direct parameters" and "derived parameter" also termed sometimes in uncertainty parlance as "multivariate parameters or variables".

Thus, just to cite an example, pressure measurement will be a direct parameter but a Mach number measured say using even a Pitot- static tube will be derived or a multivariate parameter.

Typically, for small scale scientific experiments for direct parameters multiple measurements at a single operating point may be carried out generating a large data set. This is primarily because a single measurement or a data point may be somewhat non-representative due to random error or unanticipated events. Typically, mean, median and mode are measure of central tendency of the data for a set of direct measurement systems. If say we assume, that 1 to N values of a variable 'x' are taken the arithmetic mean for the given data set is given as

$$\mu = \frac{1}{N}\sum_1^N x_i \tag{6.1}$$

The uncertainty (u_x) is then given by the standard deviation of the data set calculated using the following relation:

$$u_x = \sqrt{\frac{1}{N}\sum_1^N (x_i - \mu)^2} \tag{6.2}$$

Further, if we also consider that the measurement errors are random variables that follow probability distribution described by a probability density function, $p(x)$. Thus, in a general case uncertainty is given as

$$u_x = \sqrt{\int_{-\infty}^{\infty} (x - \mu)^2 p(x) dx} \tag{6.3}$$

Eq. (6.2) may be considered to be a special case of Eq. (6.3). In case of chemical laser systems, which are inherently complex, it is more appropriate to identify the various sources of error in the measurement process at the outset. These then may be combined to characterize the uncertainty associated in measurement of a specific physical variable. This leads us to a discussion on the sources of uncertainty in measurement. Although, in most texts this is a topic discussed subsequent to an uncertainty model but it is our understanding that knowing the various sources of error would enable the reader to appreciate the specifics of uncertainty model in a better manner.

6.1.2 Sources of Error

Measurement process errors are the basic elements of uncertainty analysis. Once these error sources have been identified, then uncertainty estimates for these errors can be developed. Some of the key errors most commonly associated with the measurement unit an applicable in chemical lasers are briefly discussed below. The reader must be mindful that these are by no means a comprehensive list and their may be other error sources as well subject to the unit being used.

a) Attribute Bias: It is a systematic error that persists from measurement to measurement [3]. For example, if there is a zero error it will persist in all measurements made by that unit. It may also occur due to bias in the reference signal used to calibrate the unit.

b) Repeatability: It is a random error which manifests itself as the variation in measured value over several measurements for given operating state of the system.

c) Resolution Error: The smallest discernible value indicated by measuring unit is termed as resolution. Resolution error is defined as the difference between the indicated value and the sensed value.

d) Operator Bias: This may be introduced due the errors made by the operator or the person making the measurement. A given measurement information may be perceived differently by different operators introducing a bias. Also, to an extent human behavior is random. Hence, the systematic part of the human error is included in operator bias whereas the random part is included in repeatability.

e) Environmental Factors Error: Environmental changes such as temperature, humidity, vibration or stray emf may also lead to introduction of errors. Consider an example of say an RTD which is based on changes in resistance on account of change in temperature. The reference resistance may change with change in temperature of environment.

f) Computation Error: Data processing errors resulting from computation, round off, truncation, numerical interpolation of values or curve fitting are categorized under this type of error. In fact all errors involved in transport of signal from the measurement unit to the data acquisition system and onwards to its display including the computation involved may be clubbed under this error type.

6.1.3 Model Generation

The prime concepts of uncertainty estimation based on sensitivity analysis for multivariate or derived parameters are well known, however, they are briefly summarized below. Any derived parameter, G, can be expressed as a function of measured fundamental quantities, $\varphi_1 \ldots \varphi_N$.

$$G = f(\varphi_1 \ldots \varphi_N) \tag{6.4}$$

There is nominal uncertainty in each of the direct or fundamental measured quantities such that

$$\varphi_i = \varphi_{i\,true} + u_{\varphi_i} = \varphi_{i\,true} + \delta\varphi_i \tag{6.5}$$

The sensitivity coefficients [4] for the derived variable G is given as

$$c_{\varphi_i} = \left(\frac{\partial G}{\partial \varphi_i}\right) \tag{6.6}$$

The component of uncertainty in G due to uncertainty in φ_i is then expressed as

$$u_{G_i} = c_{\varphi_i} u_{\varphi_i} = \left(\frac{\partial G}{\partial \varphi_i}\right)\delta\varphi_i \tag{6.6}$$

Then, provided that individual fundamental quantities (φ) are independent and normally distributed, the combined uncertainty of G is the expressed as Root Sum Square (RSS) of individual uncertainty components.

$$u_G = \sqrt{\sum_{i=1}^{N} (c_{\varphi_i} u_{\varphi_i})^2} \qquad (6.7)$$

The present work, mostly utilizes measurement of fundamental or direct measurement quantities of pressure, temperature, photo signal for system analysis. Thus, uncertainty in the measurement of these physical quantities are comprised of the *sensor errors*, the ones introduced by the sensing unit and *computation errors* introduced by the data processing, data acquisition and display system. Thus, uncertainty for these direct variables *(J)* is again represented as Root Sum Square (RSS) of these above enumerated errors *(ε_i)*.

$$u_J = \sqrt{\sum_{i=1}^{N} (\varepsilon_i)^2} \qquad (6.8)$$

6.2 Uncertainty Estimates

In this session, typical examples of uncertainty estimates for both direct and multivariate parameters are presented for better comprehension of the manner of determination. The direct parameters are taken up foremost followed by multivariate parameters since the estimates for the former influence the computation of uncertainties for the latter.

6.2.1 Direct Parameter Measurement

Consider a typical example of pressure measurement to illustrate the estimation of uncertainty in case of direct parameter measurement.

In case of chemical gas lasers both positive and vacuum pressure measurements are to be made as accurately as practically possible.

The errors are primarily attributed to the ones associated with the basic sensor unit supplied by the manufacturer and computation error which may be split into sensor power supply, signal amplifier and display unit etc.

The chain of pressure measurement is given in Table (6.1):

Table 6.1. Measurement chain with corresponding error values.

Measurement chain	Error
Positive pressure sensor	0.20 %
Vacuum pressure sensor	0.20 %
Error for 95 % reliability (1.96 σ)	0.12 %
Sensor power supply	0.05 %
Amplifier	0.05 %
Display unit or computer	0.05 %

The overall uncertainty in measurement of pressure is calculated by using root sum square [5] of the as follows

$$u_p = \sqrt{(0.2)^2 + (0.12)^2 + (0.05)^2 + (0.05)^2 + (0.05)^2} = 0.23\%$$

A similar calculation may be done for the determination of uncertainty in temperature measurement, with sensors available with similar accuracy the uncertainty works out to be of the same order.

In case of photo signals the error chain is the same, however, the error in the measurement made by the photo detector may be nearly 0.05 % by considering that the signal measured is much larger than the detector noise. The error for 95 % reliability in the measured value is expected to be of the order of 0.03 %. Assuming, no other errors with stand alone units, the estimated uncertainty is 0.058 %.

6.2.2 Derived (Multivariate) Parameter Measurement

The uncertainty analysis of derived or multivariate parameters is more involved as has been discussed in the section on model generation. It incorporates the uncertainty induced in the parameter due to uncertainty associated with various measured direct variables such as pressure, temperature, length or photo signal.

As has already been discussed in earlier chapters, chemical lasers involve measurement of several derived parameters for their characterization such as Mach number, specie concentration, small signal gain etc. In this section a few key examples of these derived parameters have been illustrated to develop the necessary understanding for their treatment.

It would be relevant to mention here that although we have mostly treated gas flow rate measurement as direct variable since it scales directly with pressure but for the purpose of uncertainty analysis it will fall under multivariate or derived parameter.

6.2.2.1 Mach Number

The easiest manner of Mach number determination is by employing Pitot- static tube for measuring Pitot and static pressure inside the test section. The uncertainty in the fundamental measured variable that is pressure occurs due to resolution, accuracy and calibration of the measuring pressure transducer and the associated uncertainties in the data processing equipment as discussed in the previous section on uncertainties in direct measurement.

The equation used for estimation of the Mach number is

$$\frac{P_{pt}}{P_{st}} = \frac{\left(\dfrac{\gamma+1}{2}M^2\right)^{\frac{\gamma}{\gamma-1}}}{\left(\dfrac{2\gamma}{\gamma+1}M^2 - \dfrac{\gamma-1}{\gamma+1}\right)^{\frac{1}{\gamma-1}}} \qquad (6.9)$$

By definition [6]

$$u_M = \sqrt{\left(\frac{\partial M}{\partial P_{st}}\delta P_{st}\right)^2 + \left(\frac{\partial M}{\partial P_{pt}}\delta P_{pt}\right)^2} \qquad (6.10)$$

$$\frac{\partial M}{\partial P_{st}} = -\frac{M\{2\gamma M^2 - \gamma + 1\}}{2\gamma P_{st}\{2M^2 - 1\}} \qquad (6.11)$$

$$\frac{\partial M}{\partial P_{pt}} = \frac{M\left\{2\gamma M^2 - \gamma + 1\right\}}{2\gamma P_{pt}\left\{2M^2 - 1\right\}} \qquad (6.12)$$

It is important to note that the uncertainty here includes only the pressure measurement uncertainty and not the errors in interpolation and probe interference. A suitable value of interpolation error may be introduced taking into account interpolation accuracy.

Thus, considering typical measured static and Pitot pressure values for supersonic COIL the uncertainty in Mach number may be computed as below.

Static pressure (P_{st})	$= 3$ torr $\pm 0.2\%$
Pitot pressure (P_{pt})	$= 11$ torr $\pm 0.2\%$
Specific Heat Ratio (γ)	$= 1.4$
Calculated Mach number (M)	$= 1.58$
Uncertainty (u_M)	$= 0.06\%$

Hence, it is evident that the uncertainty in computation of Mach number is much smaller than the uncertainty in measurement of the direct parameters from which it is computed and is to an extent also dependent on the magnitude of the mean value being measured.

6.2.2.2 Small Signal Gain

It has already been mentioned in earlier chapters that small signal gain in case of chemical lasers is one of the key parameters. It signifies the extent of population inversion inside laser cavity and directly influences the magnitude of extractable laser power. Most widely used method employed is measurement of amplification of seed laser probe beam as has already been discussed in Chapter 3. The computation of uncertainty in the measured small signal gain is enunciated below. Small signal gain is calculated employing the relation,

$$g = \frac{1}{l}\ln\left(\frac{I_v}{I_0}\right) \qquad (6.13)$$

Fundamentally,

$$u_g = \sqrt{\left(\frac{\partial g}{\partial l} \delta l\right)^2 + \left(\frac{\partial g}{\partial I_o} \delta I_o\right)^2 + \left(\frac{\partial g}{\partial I_v} \delta I_v\right)^2} \qquad (6.14)$$

$$\frac{\partial g}{\partial L} = -\frac{1}{l^2} \ln\left(\frac{I_v}{I_o}\right) \qquad (6.15)$$

$$\frac{\partial g}{\partial I_o} = -\frac{1}{l}\frac{1}{(I_o)} \qquad (6.16)$$

$$\frac{\partial g}{\partial I_v} = \frac{1}{l}\left(\frac{1}{I_v}\right) \qquad (6.17)$$

It is important to note that here only the uncertainty in intensity and length measurement are included. A suitable estimate of data processing computation error may also be included as an additional term in Eq. (6.14).

Thus, considering typical measured incident (I_o), amplified light intensity (I_v), gain length (l) and their associated uncertainty for a specific case of supersonic COIL the uncertainty in small signal gain may be computed as below.

Incident light intensity (I_o) = 4.65 mV ± 0.06 %

Amplified light intensity (I_v) = 4.77 mV ± 0.06 %

Gain Length (l) = 7.5 cm ± 0.25 %

Calculated small signal gain (g) = 0.34 % cm^{-1}

Uncertainty (u_g) = 3.34 %

6.2.2.3 Molar Gas Flow Rate

As has been mentioned earlier in this chapter that gas flow rates have mostly been treated in this text as direct variables as it scales directly with pressures but from view point of uncertainty it is treated as derived or multivariate parameter.

For the purpose of uncertainty estimation, the relation for molar flow rate is recounted in a slightly modified form of Eq. (2.18) to directly compute the gas molar flow rate.

$$G = \sqrt{\frac{\gamma}{R}} \frac{P_o A}{\sqrt{T_o} MW} \left[\frac{2}{\gamma+1}\right]^{\frac{\gamma+1}{2(\gamma-1)}} C_d \qquad (6.18)$$

The above equation for the purpose of uncertainty analysis for a given gas and discharge coefficient may be written as:

$$G = C \frac{P_o d^2}{\sqrt{T_o}} \qquad (6.19)$$

For, N_2 gas with specific heat ratio of 1.4, molecular weight (MW) of 0.028 kg and discharge coefficient of 0.95, flow constant (C) is computed to be 1.07. Thereafter, the uncertainty in measured molar flow   rates   is given as

$$u_G = \sqrt{\left(\frac{\partial G}{\partial P_o} \delta P_o\right)^2 + \left(\frac{\partial G}{\partial d} \delta d\right)^2 + \left(\frac{\partial G}{\partial T_o} \delta T_o\right)^2} \qquad (6.20)$$

The sensitivity coefficients are given as follows:

$$\frac{\partial G}{\partial P_o} = C \frac{d^2}{\sqrt{T_o}} \qquad (6.21)$$

$$\frac{\partial G}{\partial d} = C \frac{2 P_o d}{\sqrt{T_o}} \qquad (6.22)$$

$$\frac{\partial G}{\partial T_o} = -\frac{C}{2} \frac{P_o d^2}{T_o \sqrt{T_o}} \qquad (6.23)$$

Thus, considering typical measured gas stagnation pressure (P_o), stagnation temperature (T_o), orifice diameter (d) and their associated uncertainty for a specific case of supersonic COIL the uncertainty in the measured flow rate of nitrogen buffer gas may be computed as below:

Stagnation pressure of gas (P_o)	= 4.5 bar ± 0.2 %
Stagnation temperature of gas (T_o)	= 300 K ± 0.2 %
Orifice diameter (d)	= 4.5 mm ± 0.25 %
Flow constant (C)	= 1.07
Computed nitrogen flow rate	= 0.560 Moles- s^{-1}
Uncertainty	= 0.565 %

6.2.2.4 Specie Concentration

In case of chemical gas lasers measurement of various species is also critical for optimizing the laser power. In Chapter 3, we have discussed various methods for determining concentration of various species. In order to illustrate uncertainty estimation a specific case of iodine specie concentration is enunciated below.

The consolidated expression for iodine concentration measurement [6] carried out using optical absorption may be written in the form given below:

$$M_{I_2} = \frac{M_c}{P}\left\{\frac{kT}{\sigma L}\right\}\ln\left(\frac{I_o}{I_v}\right) \tag{6.24}$$

Thus, uncertainty in computed flow rate of iodine species may be expressed as

$$u_{M_{I_2}} = \sqrt{\left(\frac{\partial M_{I_2}}{\partial P}\delta P\right)^2 + \left(\frac{\partial M_{I_2}}{\partial T}\delta T\right)^2 + \left(\frac{\partial M_{I_2}}{\partial L}\delta L\right)^2 + \left(\frac{\partial M_{I_2}}{\partial M_c}\delta M_c\right)^2 + \left(\frac{\partial M_{I_2}}{\partial I_o}\delta I_o\right)^2 + \left(\frac{\partial M_{I_2}}{\partial I_v}\delta I_v\right)^2} \tag{6.25}$$

$$\frac{\partial M_{I_2}}{\partial P} = -\frac{M_c}{P^2}\left\{\frac{kT}{\sigma L}\right\}\ln\left(\frac{I_o}{I_v}\right) \tag{6.26}$$

$$\frac{\partial M_{I_2}}{\partial T} = \frac{M_c}{P}\left\{\frac{k}{\sigma L}\right\}\ln\left(\frac{I_o}{I_v}\right) \tag{6.27}$$

$$\frac{\partial M_{I_2}}{\partial L} = -\frac{M_c}{P}\left\{\frac{kT}{\sigma L^2}\right\}\ln\left(\frac{I_o}{I_v}\right) \tag{6.28}$$

$$\frac{\partial M_{I_2}}{\partial M_c} = \frac{1}{P}\left\{\frac{kT}{\sigma L}\right\}\ln\left(\frac{I_o}{I_v}\right) \tag{6.29}$$

$$\frac{\partial M_{I_2}}{\partial I_o} = \frac{M_c}{P}\left\{\frac{kT}{\sigma L}\right\}\frac{1}{I_o} \tag{6.30}$$

$$\frac{\partial M_{I_2}}{\partial I_o} = -\frac{M_c}{P}\left\{\frac{kT}{\sigma L}\right\}\frac{1}{I_v} \tag{6.31}$$

Thus, considering typical measured total gas pressure *(P)*, temperature *(T)*, probe cell length *(L)*, carrier gas mass flow rate *(M_c)*, incident light intensity *(I_o)*, transmitted light intensity *(I_v)* and their associated uncertainty for a specific case of supersonic COIL the uncertainty in the measured iodine flow may be computed as below:

Total gas pressure *(P)* $\quad = 150$ torr $\pm\, 0.2\,\%$

Probe cell temperature *(T)* $\quad = 353$ K $\pm\, 0.2\,\%$

Probe cell length (L) $\quad = 10$ cm $\pm\, 0.25\,\%$

Incident light intensity *(I_o)* $\quad = 6$ V $\pm\, 0.06\,\%$

Transmitted light intensity *(I_v)* $\ = 2$ V $\pm\, 0.06\,\%$

Carrier gas flow rate *(M_c)* $\quad = 1.5$ Mole-s^{-1} $\pm\, 0.56\,\%$

Computed Iodine flow rate $\quad = 0.019$ Moles-s^{-1}

Uncertainty $\qquad\qquad\qquad 0.67\,\%$

In conclusion, it would be prudent to say that a statement on uncertainty is essential to ascertain the efficacy of the measurements made in complex systems such as chemical gas lasers. Only typical examples have been illustrated but the concepts are generic and may be applied to all kinds of measurement configuration. The reader will also appreciate that each parameter measurement may be made in a multitude of ways e.g. measurement of specie concentration, Mach number etc. The overall uncertainty is also a function of the measurement methodology in terms of the equipment employed and computation or interpolation error. An example incase is Mach number measurement using Voigt profile method requires curve fitting and a more robust and accurate results than an algorithm that is not so efficient.

We hope that the above discussion would be a precursor for carrying out uncertainty estimation for various other relevant parameters pertaining to chemical lasers and also for other complex systems that the reader may happen to deal with.

References

[1] S. Castrup, Obtaining and Using Equipment Specifications, in *Proceedings of the NCSLI Workshop & Symposium*, Washington, D. C., August 2005.

[2] H. W. Coleman and W. G. Steele, Experimentation and Uncertainty Analysis for Engineers, 2nd Edition, *Wiley Interscience Publication, John Wiley & Sons, Inc.,* 1999.

[3] R. M. Gray, Probability, Random Processes, and Ergodic Properties, *Springer-Verlag,* 1987. Revised 2001 and 2006-2007 by Robert M. Gray.

[4] D. Deaver, Having Confidence in Specifications, in *Proceeding of the NCSLI Workshop and Symposium,* Salt Lake City, UT, July 2004.

[5] K. Annamalai, K. Visvanathan, V. Sirramulu, K. A. Bhaskaran, Evaluation of the performance of supersonic exhaust diffuser using scaled down models, *Experiments Thermal Fluid Science,* 17, 1998, p. 217.

[6] Mainuddin, Gaurav Singhal, R. K. Tyagi, and A. K. Maini, Diagnostics and data acquisition for chemical oxygen iodine laser, *IEEE Transactions on Instrumentation and Measurement,* 61, 6, 2012, p. 1747.

Index